Better Math in

5 Minutes a Day

Fun Math Activities for Kids and Parents on the Go

Fran Gibson

Series Editor: Mark Pennington

PRIMA PUBLISHING

3000 Lava Ridge Court • Roseville, California 95661
(800) 632-8676 • www.primalifestyles.com

© 2001 by Prima Publishing

All rights reserved. No part of this book may be reproduced or transmitted in any form or by any means, electronic or mechanical, including photocopying, recording, or by any information storage or retrieval system, without written permission from Prima Publishing, except for the inclusion of brief quotations in a review.

The 5 MINUTES A DAY logo is a trademark of Prima Communications, Inc. PRIMA PUBLISHING and colophon are trademarks of Prima Communications Inc., registered with the United States Patent and Trademark Office.

Interior illustrations by Susan Sugnet

Library of Congress Cataloging-in-Publication Data
Gibson, Fran.
    Better math in 5 minutes a day : fun math activities for kids and parents on the go / Fran Gibson.
       p. cm. — (5 minutes a day series)
    Includes index.
    ISBN 0-7615-2427-4
    1. Arithmetic—Study and teaching (Elementary) I. Title: Better math in five minutes a day. II. Title. III. Series.
QA135.6.G56 2001
372.7—dc21   2001016346

01 02 03 04 II 10 9 8 7 6 5 4 3 2 1
Printed in the United States of America

How to Order
Single copies may be ordered from Prima Publishing, 3000 Lava Ridge Court, Roseville, CA 95661; telephone (800) 632-8676, ext. 4444. Quantity discounts are also available. On your letterhead, include information concerning the intended use of the books and the number of books you wish to purchase.

Visit us online at www.primalifestyles.com

I dedicate this book to my very first math teacher and my very first angel, Peg Gibson.

I love you, Mom.

Contents

Acknowledgments vi

Introduction vii

Chapter 1: MULTIPLICATION   1

Multiplication: The Times Are Changing   2
Just the Facts, Nothing But the Basic Facts   5
The Power of Ten   13
Multiplication: Beyond the Simple Facts   17
Estimation and Rounding: I Guess I Can   20
Factors Factors Factors   24
The Greatest Common Factor of All   28

Chapter 2: DIVISION   32

Division and the Art of Being Fair   33
The Power of Ten—in Reverse   39
Daddy, Mommy, Sister, Brother … and More   42

Chapter 3: FRACTIONS   47

Fraction, Fraction, What Is a Fraction?   48
There's a Whole in This Fraction   52
Where's the One (or the Two, for That Matter)?   55
Improper Fractions: Not a Matter of Manners   60
Where, Oh Where Is That Half?   64
What's It Worth to You?   69
Equivalents: Different but the Same   72

Chapter 4: WORKING WITH FRACTIONS   75

Add a Little, Subtract a Little   76
Sensible Multiplication   83
The Rules of Multiplication   87
Division Made Easy   90
The Rules of Division   94

Chapter 5: DECIMALS   98

Decimals: Friends to Fractions   99
Decimal Draw   104
Fractions to Decimals   107
Decimals to Order   110
Adding and Subtracting Decimals   115
Multiplying and Dividing Decimals   121

Answers   126

Index   135

Acknowledgments

I would like to acknowledge some very important supports and resources that took part in helping to create this book. First, I thank my series editor, Mark Pennington, for criticisms and compliments. Thanks also go to Jamie Miller and Tara Mead at Prima for their ideas and their editing expertise. I am much appreciative of my colleagues, Chris and Tina, who were properly impressed that I was writing a book and getting paid to do it. I am forever grateful to Tom, Nancy, and Granny Pat for their guidance in my math life. I thank and admire all of my Math Matters teachers and colleagues, especially those at Howe Avenue and Kenneth Avenue Schools in Sacramento. They consistently teach me about dedication and perseverance. I am grateful to my extended family, Mom, Maggie, Mike, Red, Sylvia, Christine, Kenneth, Ellen, Paul, Henneberry, Mike, Margie, Wes, and every single Graf, who let me know how proud they were of my newest challenge. I thank all the children in my life, my son, my nieces, my nephews, students, neighbors, and friends, for always reminding me of the beauty in the smallest bits of life. But, most importantly, I thank my family, Jody and Riley, for supporting my work and for loving my heart.

Introduction

Mathematics is an essential part of life. It is the language of our daily calculations and our means to communicate many of the scientific advances that shape our world. We depend on mathematics. Why, then, given its importance, is this human tool the source of fear and frustration for both children and adults?

Educators say that people who have difficulty learning mathematics do not have number sense. Parents say that their children do not see the connections between concepts. Both groups are saying the same thing.

People who understand mathematics have solid number sense. They can see the relationships among numbers. They can estimate. They connect concepts and use what they know to solve new problems. They rely on the patterns of mathematics to help them make sense of new situations. They can see the "reasonableness" of an answer. They ask questions. They enjoy mathematics.

This book is a collection of ideas to use to build number sense with multiplication, division, fractions, and decimals. This book is not intended to be used in place of school math curriculum. Rather, it is designed to enhance and deepen intermediate-age children's understanding of mathematics.

Each chapter comprises lessons to guide you (the parent/teacher) and your child (the learner) through each concept. The Parents' Corners explain the concept. This is your chance to sort out the goals of the lesson. Parents' Corners also include compelling reasons for learning the concept and story examples to illustrate.

Teaching Tips offers ideas to help you incorporate math concepts throughout your day in a way that will enhance your child's understanding. They also provide tips for how best to proceed with the lesson. The pace for each lesson will vary; understanding the concept is always the goal.

Each lesson contains a variety of activities: At the Kitchen Table, On the Go, and On Your Own. At the Kitchen Table activities involve the parent and the child and usually require paper and pencil. Games in this category help children have fun while learning to understand concepts and developing strategies to solve problems. During the activities and games, encourage your child to explain concepts both verbally and in drawings.

On the Go activities can be done in a car, while waiting in a line, or anywhere else you happen to be and have a few minutes. These activities are designed to help your child build mental mathematical ability. The more math your child can calculate mentally, the more efficient he or she will be. In addition to these activities, ask questions to spark conversations about numbers in everyday routines.

On Your Own activities are supplementary activities that your child can do without your help. Instructions for these activities are written directly for the child and are at a child-appropriate level. But this doesn't mean you can't help. If your child wants to talk to you about the On Your Own activities, feel free to work through them.

Finally, Imagine That! and Just for Fun! items have been scattered throughout the lessons to remind both you and your child that math is fun.

This book is designed to help you and your child have fun while learning math. It is filled with story problems, strategies, real-life applications, and, most importantly, ways for today's children to understand the math that most adults learned through memorization in earlier generations.

Multiplication

In This Chapter

- Multiplication: The Times Are Changing
- Just the Facts, Nothing But the Basic Facts
- The Power of Ten
- Multiplication: Beyond the Simple Facts
- Estimation and Rounding: I Guess I Can
- Factors Factors Factors
- The Greatest Common Factor of All

MULTIPLICATION: THE TIMES ARE CHANGING

Parents' Corner

Mastering multiplication is a rite of passage for the growing mathematician. To learn the concept, children must understand and memorize a lot of new information. Your child will have the best chance of retaining this new information and using it correctly if he or she understands each step as he or she learns it.

What is multiplication, anyhow? Simply put, it is a fast way to do repeated addition. This connection to addition is important. A child could add a number repeatedly ($3 + 3 + 3$) or learn the multiplication fact (3×3) to get the same answer.

Let's see how this works. How many wheels do four bicycles have in all? You could add up 2 wheels + 2 wheels + 2 wheels + 2 wheels = 8 wheels. *Or* you could multiply 4 bikes $\times$ 2 wheels each = 8 wheels in all.

Multiplication, like addition, is flexible. Both 4×2 *and* $2 \times 4 = 8$. This property is called the *commutative property of multiplication*.

When you and your child work with mathematics, use the proper terms whenever you can. For multiplication, the terms are *factor* and *product*: Factor $\times$ Factor = Product. Using the proper terms will help your child learn and become comfortable with the vocabulary of multiplication.

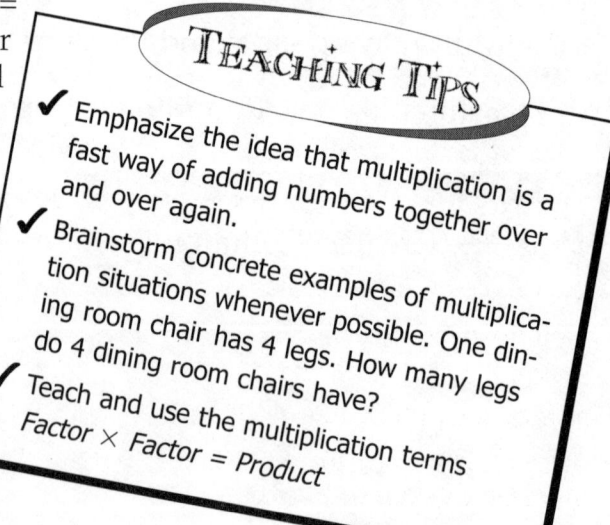

Teaching Tips

- ✓ Emphasize the idea that multiplication is a fast way of adding numbers together over and over again.
- ✓ Brainstorm concrete examples of multiplication situations whenever possible. One dining room chair has 4 legs. How many legs do 4 dining room chairs have?
- ✓ Teach and use the multiplication terms Factor $\times$ Factor = Product

At The Kitchen Table

Dice Multiplice

Here is a good game to teach you the meaning of multiplication. Play it for a few minutes at a time until you understand how multiplication works.

To play, you need 2 dice, 6 bowls, and a pile of stuff, such as pennies, beans, toothpicks, or buttons. Roll the first die to find out how many bowls to put out. The bowls show the number of groups you need. Roll the second die to find out how many things to put in each group (bowl). For example, if you roll a 4 and then a 3, set out 4 bowls with 3 toothpicks in each. If you count up the number of toothpicks, you'll see that $4 \times 3 = 12$.

Say the multiplication equation while making the groups: "Four times three equals twelve." The more you repeat it, the better you'll remember it.

On Your Own

Skip-Counting Serenade

Skip-count to yourself whenever you can. You can even make up a chant or put the numbers to music if it helps you. Sing your skip-count song when you take a bath. Shout out your numbers as you kick a soccer ball around. Whisper numbers as you walk your dog. Who knows? Your dog might even learn a few multiplication facts!

> **Imagine That!**
> Did you know that *arithmophobia* is the fear of numbers? After you finish the activities in this book, you won't suffer from this fear!

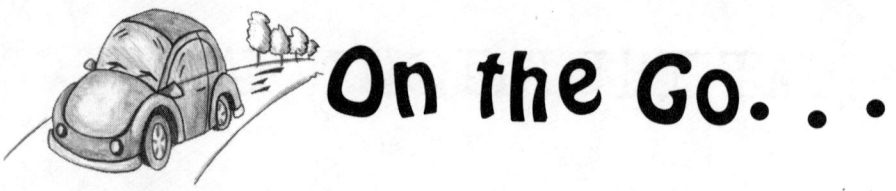
On the Go...

SKIP-COUNTING CALISTHENICS

This activity comes in handy for teaching multiplication facts, especially for children just beginning to learn the concept. When your child does counting calisthenics, it exercises his or her mind and makes it stronger, just as jumping jacks exercise and strengthen the body. These exercises can be done just about anywhere. As you walk to the park, skip-count by 4s. As you search the grocery aisles for orange juice and milk, skip-count by 8s. Skip-count by 6s while waiting at a traffic light. Simply give your child a number to skip-count by. For example:

You say: Skip-count by 2.
Your child says: 2, 4, 6, 8, 10, 12, 14 . . .

After your child really gets going, give one number to skip-count by and another for the number of times to skip-count. For example:

You say: Skip-count by 6 four times.
Your child says: 6, 12, 18, 24

JUST THE FACTS, NOTHING BUT THE BASIC FACTS

Parents' Corner

Your child will need both time and effort to learn the facts. Be patient. Even if your child is good at memorizing facts, make sure he or she knows what those facts mean. Memorization works well, but only if it is accompanied by an understanding of what is being memorized.

Encourage your child to use the strategies in At the Kitchen Table: Strategies Strategies Strategies. But beware: Some of these strategies may seem complicated at first. In fact, it might take just as long to learn the strategies as it takes to learn the facts themselves. Even so, remember that strategies are often valuable because they can help your child *understand* the facts being memorized.

Doing basic computations mentally rather than on paper or by calculator is mathematically efficient. The more math facts your child can store in his or her head, the better your child will be at recalling those facts quickly when it's time to use them. As your child learns to *compute*, or "crunch," numbers mentally, he or she strengthens the understanding of how numbers relate to one another. In turn, number sense grows.

If your child can use the strategies suggested in this chapter with one-digit numbers, he or she will find it easier to work with bigger numbers later. Emphasize that these facts are necessary, not only for school math work but also for real-life situations.

Teaching Tips

✓ Help your child visualize math by saying such things as, What does 7 × 3 mean? Draw a picture of what is happening. What are some real-life examples of 7 × 3?

✓ Tell your child how you learned your multiplication facts, or times tables, when you were a child.

✓ Keep emphasizing that mathematics will be easier in the future if your child uses strategies and memorizes facts.

AT THE KITCHEN TABLE

STRATEGIES STRATEGIES STRATEGIES

If you work hard to exercise your brain, not only will you memorize the facts, but you will also learn how to figure out in your head those tough facts that you haven't memorized. These facts will come in handy if you are ever stranded without a calculator—maybe on a deserted island, at a grocery store, or in the middle of a math test!

The strategies below have been put together to help you learn the facts. "Strategy" is just a fancy word for a plan that you follow to get where you want to go. In the case of mathematics, you want to get from the problem to the answer. You also want to remember how you got to the answer. After reading each strategy, use it to find the answers to the problems following it. After you've mastered each strategy, have your parent give you other problems using the same strategies.

× 0   Anything times 0 is 0. This is called the *Zero Property of Multiplication.* Try these:

$7 \times 0 =$ _____ $43 \times 0 =$ _____ $3{,}924 \times 0 =$ _____

× 1   The number 1 follows the *Identity Property of Multiplication.* This means that any number multiplied by 1—no matter how high the number—is always the same number. In other words, the number being multiplied by one keeps its own identity. The number 1 cannot do anything to it. Try these:

$7 \times 1 =$ _____ $43 \times 1 =$ _____ $3{,}924 \times 1 =$ _____

× 2   Times 2 means to double. To understand this strategy, think of double scoops of ice cream. Double 1 scoop of chocolate chip ice cream to get 2 scoops. Double 2 scoops of strawberry ice cream to get 4 scoops. Practice doubling numbers to learn your times 2s. Try these:

4 × 2 = _____   6 × 2 = _____   8 × 2 = _____

× 3   Just like times 2, times 3 comes with a quick trick. If you can triple numbers, you can multiply by 3. Practice a few triples: 7 × 3, or triple 7 (7 + 7 + 7); both equal 21. This trick works for all numbers. 42 × 3 is the same as triple 42 (42 + 42 + 42). Both equal 126. Try these:

4 × 3 = _____   5 × 3 = _____   8 × 3 = _____

× 4   Times 4 takes a little practice. It requires doubling twice. For example, 6 × 4 is the same as double-6 plus double-6—or four 6s. 6 + 6 = 12 (one double). Double 6 again for another 12. Then add the two doubles together (12 + 12) to get 24. Therefore, 6 × 4 = 24. Try these:

4 × 4 = _____   5 × 4 = _____   8 × 4 = _____

× 5   Two simple strategies work for times 5. One is to skip-count by 5s for a quick answer. The other works well if you can quickly cut numbers in half. For this second strategy, simply multiply by 10 and cut that number in half. (See the Power of Ten lesson later in this chapter if you need help multiplying by 10.) This strategy works because 5 is half of 10. Take 6 × 5. 6 × 10 = 60. Half of 60 is 30, so 6 × 5 = 30. Try these:

4 × 5 = _____   6 × 5 = _____   7 × 5 = _____

× 6   This strategy combines two other strategies. It's a bit complicated, but once you understand it, you'll find it easier to learn the facts. Since 3 × 2 = 6, you can use the × 3 and × 2 strategies to find × 6. Triple a number (× 3) and then double the sum of the triple (× 2) to get times 6. Take 7 × 6. First, triple 7 (7 + 7 + 7) to get 21. Then double the sum of the triple (21 + 21) to get 42. 7 × 6 = 42! Try these:

3 × 6 = _____   4 × 6 = _____   8 × 6 = _____

× 7   This is tough. We'll come back to it after we've looked at all the others.

× 8   This strategy is a double bonanza. To multiply by 8, double, double, and double again. This works because 2 × 2 × 2 = 8, so any number that you double, double, and double again will be multiplied by 8. Let's look at 7 × 8. Double-7 = 14; double-14 = 28; double-28 = 56. 7 × 8 = 56! Try these:

4 × 8 = _____   6 × 8 = _____   9 × 8 = _____

× 9   To multiply by 9, triple and triple again. This makes sense, because 3 × 3 = 9. Try 7 × 9. First triple 7. You get 21. Now triple 21. You get 63. Yes, 7 × 9 = 63. Or, you could multiply your other number—not 9—by 10 and then subtract one of that number. To find 7 × 9, you would multiply 7 × 10 to get 70, then subtract 7 to get 63. This works because nine 7s is the same as ten 7s minus one 7. (*Tip:* For all multiples of 9, the digits add up to 9. So 7 × 9 = 63, and 6 + 3 = 9; 6 × 9 = 54, and 5 + 4 = 9; and so on. This is a good way to check your work!)

4 × 9 = _____   8 × 9 = _____   9 × 9 = _____

× 7 revisited   Now that you have learned the strategies for 0 through 9, you have only one 7 to learn: 7 × 7. This is easy if your favorite football team happens to be the answer to this equation! (49)

THE NUMBER CRUNCHER

A t-chart is a tool that mathematicians use to organize information. In this game the t-chart is a number-crunching machine that will help you practice and solve problems. Find someone to play with (a parent, for example). Think of a multiplication fact—such as × 4—but don't tell your opponent your fact. Draw a t-chart on a piece of paper (see illustration). Have your opponent give you a number to put in the IN column of the t-chart machine. In the sample here, the opponent has suggested first 3 and then 5.

IN	OUT
3	___
5	___

Show your opponent what happens to the number after you use your fact on it. In this example, the number 3 from the IN column results in 12 after you use your fact on it. So you would write 12 in the OUT column. Have your opponent continue giving you numbers for the IN column until he or she figures out the rule. Then switch roles. Now *you* have to figure out what the function rule is. Good luck!

> **IMAGINE THAT!**
> Multiply 37,037 by any single number (1–9) and then multiply that number by 3. Every digit in the answer will be the same as that first single number.

On Your Own

MULTIPLICATION CHART

Use Strategies, Strategies, Strategies from At the Kitchen Table or the refresher information below as you work through the multiplication fact chart. Cross out any facts from the chart that you know or that you can figure out quickly by using the strategies.

1. Find all the × 1. Review the Identity Property if that helps. Cross out the facts you know.

2. × 2 is just doubling. Cross out the ones you know.

3. × 3 is tripling. Cross out the facts if you can figure them out quickly.

4. Double-double to get rid of × 4.

5. For × 5, remember your two options: Skip-count by 5s or multiply by 10 and cut in half. Cross out the × 5s if they are easy for you.

6. Triple and double to get × 6. Cross out the × 6s if you can do them quickly.

7. Double-double-double to get rid of × 8.

8. Triple-triple, and you've made it through × 9.

Multiplication

X	1	2	3	4	5	6	7	8	9
1	1	2	3	4	5	6	7	8	9
2	2	4	6	8	10	12	14	16	18
3	3	6	9	12	15	18	21	24	27
4	4	8	12	16	20	24	28	32	36
5	5	10	15	20	25	30	35	40	45
6	6	12	18	24	30	36	42	48	54
7	7	14	21	28	35	42	49	56	63
8	8	16	24	32	40	48	56	64	72
9	9	18	27	36	45	54	63	72	81

Look at the chart now. What do you still have to memorize? 7 × 7. Remember the football team! Go 49ers!

If you've left other numbers uncrossed, practice those next. With fewer facts to worry about, you'll have them learned in no time!

Just for Fun!

Question: How can you divide six potatoes evenly among 20 people?

Answer: Boil and mash them!

On the Go...

MEMORY LANE

The best way to memorize something is to hear it again and again. Think of songs that you've heard on the radio so many times that you know them backward and forward. Did you set out to study them and learn each word? No. You just heard them over and over, and you probably enjoy the song. You can use these two ingredients—repetition and fun—to help your child learn the multiplication facts.

Quiz your child on the facts that he or she is learning. Your child *must* repeat back the entire equation. If you say 6×7, your child *must* answer, $6 \times 7 = 42$. Use any free time to practice these facts—in the car, in the grocery line, in the waiting room at the dentist's office . . . anywhere.

THE POWER OF TEN

Parents' Corner

After your child becomes more comfortable with basic multiplication facts, you are ready to journey together through the multiplication universe. The first transition you'll make is from basic fact to "basic-fact-in-hiding." In other words, the basic fact is hidden in the bigger fact. For example, your child has learned $4 \times 8 = 32$; the natural progression is $40 \times 8 = 320$. Children, however, often do not see the relationships between the basic facts and the bigger multiplication problems. You might need to emphasize this connection. A good way to start is with powers of 10.

A number is called a power of 10 when it can be expressed by multiplying only 10s. For example, $1,000 = 10 \times 10 \times 10$—or, as mathematicians say, 10^3. $10^3 (1,000)$ has three 0s. $1,000,000$ can also be expressed by multiplying only tens: $10 \times 10 \times 10 \times 10 \times 10 \times 10 = 1,000,000$. This number, 10^6, has six 0s.

Keep in mind that multiplying a number by a whole number makes it bigger. When you multiply by 2, your answer is 2 times bigger than what you started with. When you multiply by 10, your answer is 10 times bigger!

Teaching Tips

- ✓ Point out to your child the pattern in multiplying by 10: Whenever you multiply by 10, you add a zero. Start simple (10×6, 10×9) and work to harder equations (10×14, 10×22).
- ✓ After your child really understands the idea of making something 10 times larger, move up to 100 times and 1,000 times. These new concepts should then come easily.

AT THE KITCHEN TABLE

THE TRICK OF TENS

To multiply by 10, simply add a 0 to the right of the other number in the equation. What happens if you multiply by the powers of 10, 100, 1,000, or even 1,000,000? Have an adult help you figure out the answers to these equations.

1. 6 × 100 = ____ **2.** 23 × 100 = ____ **3.** 82 × 100 = ____

What is happening to the numbers? What trick have you discovered? Try these:

4. 8 × 1,000 = ____ **5.** 9 × 1,000,000 = ____ **6.** 12 × 10,000 = ____

Practice multiplying by powers of 10 often.

MULTIPLICATION CONCENTRATION

Write the following multiplication problems and numbers on small pieces of paper (index cards or old business cards will work great). Put only one multiplication problem or answer on each piece of paper.

40 × 8	320	800 × 4	3,200
90 × 200	18,000	9 × 20	180
2,000 × 3	6,000	20 × 30	600
40 × 200	8,000	400 × 2	800
50 × 50	2,500	500 × 50	25,000

Find someone to play the game with you. Place all the cards face down in rows of three or four. Whoever goes first turns over two cards. If the multiplication problem and the answer match, that person keeps the cards and takes another turn. If the cards do not match, place them face down exactly where they were found so the next player can take a turn. Keep playing until all cards are matched. Count who has the most matches in the end. By remembering where the card with each problem is in your spread, you'll help yourself remember the multiplication facts.

You can keep playing this game with the same facts, or you can make up some new facts. Have each player come up with five or six equations and answers. The more cards you add to the pile, the tougher the game becomes!

Caution: *Beware of the zeros. Use what you know about multiplying by zeros to make sure you have the right answer.*

On the Go...

GROCERY BONANZA

While in the grocery store, find items in packaged quantities and test your child's power of ten multiplication skills. If you bought 10 packages of 6 cupcakes, how many cupcakes would you have? (60 cupcakes) How many would you have if you bought 100 packages? (600 cupcakes) How many donuts would you have if you bought 10 boxes of 8 donuts each? (80 donuts) How many sodas are there in 100 cases of 24 sodas each? (2,400 sodas) What if you bought 10 packages of 36 cookies each? (360 cookies) What about 1,000 packages with 5 sticks of gum each? (5,000 sticks of gum) Have fun with this. Some of the numbers could bring giggles as you imagine the vast quantities!

On Your Own

Fancy Multiplications

You know the patterns of multiplying by powers of 10. Now use what you know to do some fancy multiplying. Try these multiplication patterns:

1. 8 × 2 = _____   **2.** 8 × 20 = _____   **3.** 8 × 200 = _____

Challenge: **4.** 80 × 20 = _____   **5.** 800 × 200 = _____

6. 9 × 3 = _____   **7.** 90 × 3 = _____   **8.** 900 × 3 = _____

Challenge: **9.** 90 × 30 = _____   **10.** 900 × 300 = _____

11. 7 × 4 = _____   **12.** 70 × 4 = _____   **13.** 700 × 4 = _____

Challenge: **14.** 70 × 40 = _____   **15.** 7,000 × 40 = _____

Imagine That! For every one person on earth, there are 200,000 ants.

Work on these patterns so that multiplying by any multiple of 10, 100, or beyond becomes easy.

MULTIPLICATION: BEYOND THE SIMPLE FACTS

Parents' Corner

Multiplication is the beginning of many great adventures. A child who knows the basic multiplication facts and who can multiply any number by a power of 10 is ready for anything, even to multiply two-digit numbers mentally. In fact, after developing a firm understanding of the basic multiplication facts or learning strategies to remember them quickly, your child will be able to connect to all other multiplication problems consistently.

Multiplication equations can be taken apart into simpler equations. For example, think about 23 × 6. Leave the paper and pencil alone for a minute; you can do this mentally. 23 is the same as 20 + 3. So 23 × 6 can be broken into two smaller equations: 20 × 6 and 3 × 6. You know 20 × 6 = 120 (an easy way to figure this out is to double 10 × 6) and 3 × 6 = 18. Simply add the products 120 + 18 to get your answer: 23 × 6 = 138!

Here's another example. Suppose your child is faced with the problem 23 × 41. 23 is the same as 20 + 3. 41 is the same as 40 + 1. So

$$\begin{array}{r} 23 \\ \times\ 41 \\ \hline \end{array} = \begin{array}{r} 20 + 3 \\ \times\ 40 + 1 \\ \hline \end{array}$$

The facts your child has to compute are 1 × 3, 1 × 20, 40 × 3, and 40 × 20. Add up all the products to get the answer! 3 + 20 + 120 + 800 = 23 × 41 = 943.

TEACHING TIPS

✓ Begin by taking some simple multiplication problems apart into two small equations. 12 × 5 = 10 × 5 plus 2 × 5; 14 × 3 = 10 × 3 plus 4 × 3.

✓ Allow your child to use paper and pencil at first. Encourage mental computations as he or she improves.

AT THE KITCHEN TABLE

MULTIPLICATION MYSTERY

To play this game, you need a partner, a die, and nine cards with the numbers 1 through 9 written on them (one digit per card).

If you are the first player, shuffle the cards and put them facedown in a pile. Pick two cards from the top of the pile, and then roll the die. Place the cards so that you get a two-digit number. Then multiply that two-digit number by the number on the die to create a multiplication mystery.

To solve the problem, find the two small equations hiding in the big problem. Let's say you drew a 7 and a 3, and the die landed on 4. If you made 37 with the cards, then the smaller equations are 30×4 and 7×4. You know that $30 \times 4 = 120$ (an easy way to find this is $10 \times 4 = 40$; $40 \times 3 = 120$), and $7 \times 4 = 28$ (remember the double bonanza!). Add the two numbers together to get the answer: $120 + 28 = 148$.

After each turn, put the cards at the bottom of the pile. After you go through the pile once, be sure to shuffle the cards!

On Your Own

FAVORITE DIGITS

Choose three of your favorite two-digit numbers. You might pick your birthday, the number of your favorite baseball player, and the score on your last math test (unless you got 100!). Let's say you choose:

12, 36, 82

Now choose a single-digit number. Be kind to yourself at first, and choose a low number. How about 4? Multiply your three favorite numbers by 4—but not all at once! One at a time, look for the two small equations in each. Here's a look:

$$12 \times 4 = 10 \times 4 + 2 \times 4$$
$$= 40 + 8$$
$$= 48$$

IMAGINE THAT!
If you multiply 526,315,789,473,684,210 by *any* number, you will always find the original number in the result.

You did it! Your numbers are thankful; your brain is thankful. Try another.

After you finish multiplying your favorite numbers by 4, take a break, clean out your sock drawer, and then pick a new single-digit number and start all over again.

On the Go...

Mental Multiplication

This is a great game for car rides, and you don't need any paper or pencils. Take three numbers off of a license plate—for example, 2, 9, and 4. Put them together to make a two-digit number and a separate one-digit number. Use the two numbers to create a multiplication equation that will stump your child. In this example, you might have 92×4. Help your child remember the trick: Find the two smaller equations, 90×4 and 2×4. $360 + 8 = 368$ (be careful adding). Find another car and three more numbers.

ESTIMATION AND ROUNDING: I GUESS I CAN

Parents' Corner

How many times has your child given you an outrageous answer to a problem and then, when you ask where that answer came from, replied, "I don't know. It's just what I got!" The child who does that hasn't learned about estimation. Estimation is the key to number sense. Someone who can estimate an answer before calculating it exactly will always know whether the answer is reasonable. The best mathematicians, in fact, are great estimators.

There are two common ways to estimate: by rounding or by front-end estimating:

> **TEACHING TIPS**
> - ✓ Decide with your child which method of estimation he or she thinks is easier and use that method for these activities.
> - ✓ Make estimation routine whenever you do mathematics: Do you think 8 × 9 is more or less than 100? What makes you think that?
> - ✓ Help your child see the power of estimating and rounding in everyday life: If we buy 6 cans of juice for $1.23 each, about how much money will we need? Will $10.00 be enough?

For *rounding*, use the closest multiple of 10 for a two-digit number, 100 for a three-digit number, and so on. If the number ends in 5, round up (for example, 45 rounds up to 50).

$$45 \times 31$$

45 rounds up to 50   31 rounds down to 30

$$50 \times 30 = 1{,}500$$

Your answer should be around 1,500.

For *front-end estimating*, round down to the place value of the first digit:

$$45 \times 31$$
$$40 \times 30 = 1{,}200$$

Your answer should be around 1,200

45 × 31 = 1,395, which is between 1,200 and 1,500! The precise number is often important—but, estimation and rounding can help your child avoid mistakes.

AT THE KITCHEN TABLE

PIGS AND CHICKENS

For this game, you need 1 die, 2 pencils, and some paper. Invite either a parent or friend to play along.

Decide whether you want to play the Rookie version or the Veteran version. You might start with the Rookie version and move up to Veteran when you feel ready.

Rookie Version

The aim of the Rookie version is to get a score as close to 1,000 as possible. Before playing, roll the die twice to get a two-digit number. If you roll a 7 and a 4, your number will be 74. Write this two-digit number on a piece of paper. This number will be the multiplicand (the factor you will multiply each time you roll the die) for you and your opponent throughout the game.

Take turns rolling the die. Multiply the single-digit number you roll by the multiplicand. The product of the multiplication is your score. If you roll a 3, multiply it by 74 to get 222. Record this number on your scorecard, and let your opponent take a turn. Each player keeps his or her own score.

On your next turn, you have to make a big decision. Should you roll again? You want to get as close to 1,000 as you can without going over. If you go over, you lose! Use your estimation skills to figure out roughly how many more points you can get. Will a 6 on the die put you over the top? A 5? Don't take time to figure out every possibility. Simply estimate whether you are likely to stay under the target with another roll.

In the example above, you could get a product of about 700 and still stay under the target. If you know that 6 (the highest number you can get from the die) times 74 is about 420, then you can safely roll again. If you decide to go for it, say

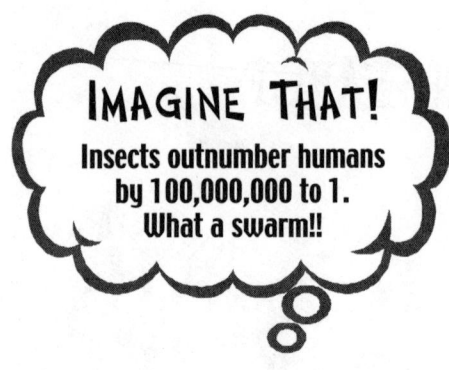

IMAGINE THAT!
Insects outnumber humans by 100,000,000 to 1. What a swarm!!

"pig" and roll the die. If you decide to stop, say "chicken." Once you say "chicken," you cannot change your mind and try again later. What a big decision!

After you finish your turn, your parent will have to make the same decision. The game continues until someone goes over the target number or until both players have become chickens. If both are chickens, the player closest to the target number wins.

Veteran Version

For the Veteran version, use 2,000 (or more) as your target. Before playing, roll the die three times to get a three-digit number to use as the multiplicand.

On Your Own

Adventure Math

Getting to school can be a great adventure. Today, put a little math into your adventure. On your way to school, keep your eyes open for license plates. Look for plates that have at least four numbers on them. As you travel to school, estimate the answer to the first number multiplied by the three-digit number formed by the remaining numbers on the license plate.

4 × 216

216 is about 200, so 4 × 200 is 800. The estimate is 800.

GROCERY STORE MANIA

On your next shopping trip, drill your child on estimation skills. In the fruit section, for example, call out "Bananas are 2 pounds for 99 cents! About how much will 6 pounds cost?" Your child will estimate:

    2 lb = $0.99, or about $1.00
    6 lb = 2 lb + 2 lb + 2 lb
    6 lb = about $1.00 + $1.00 + $1.00 = about $3.00
    6 lb will cost about $3.00

    Continue to call out the item, price, and number of items to calculate. As another example, "Yogurt, $2.59 a six-pack! Let's get three six-packs!" Your child could round $2.59 to $3.00, following the rule of rounding half a dollar or more upward. $3.00 × 3 = $9.00. Or your child could round $2.59 down to $2.50. The difference between the real number and the rounded number is only 9 cents on each 6-pack of yogurt. $2.50 × 3 = $7.50. Or, a child who is really good with numbers could round $2.59 to $2.60, multiply $2.60 by 3 or double $2.60, and then add $2.60 again. The result would be

    $2.60 × 3   = ($2.00 × 3) + ($0.60 × 3)

                     = $6.00 + $1.80

                     = $7.80

or $2.60 + $2.60 = $5.20 + $2.60 = $7.80

    Whatever strategy your child uses, encourage him or her to do as much of the work as possible in the head. Discuss with your child how he or she arrived at the answer.

FACTORS FACTORS FACTORS

Parents' Corner

As children gain a stronger understanding of multiplication, they are able to pull apart numbers into factors and eventually to find the greatest common factor of two numbers. This will help them in working with equivalent fractions and with adding and subtracting fractions.

Numbers that have only two factors—1 and the number itself—are called *prime numbers.* That is, you can divide the number only by 1 or the number itself to get a whole number for the answer. The number 11 is an example of a prime number. Its only factors are 1 and 11. Numbers that have more than two factors are called *composite numbers.* Most composite numbers have an even number of factors. For example, 24 has the factors 1, 2, 3, 4, 6, 8, 12, and 24. A few composite numbers, called *square numbers,* have an odd number of factors. The square number 16, for example, has the factors 1, 2, 4, 8, and 16. The number 1 is unique: It is neither prime nor composite. It does not have exactly two factors, but neither does it have more than two factors. One is the loneliest number.

Make sure your child remembers what a factor is. Remind him or her of the terms learned early in this chapter:

Factor × Factor = Product.

> **Teaching Tips**
>
> ✓ Help your child investigate each number to find all of its factors. Allow time to make sure that your child understands the ideas.
>
> ✓ When working with factors, refer to the Divisibility Rules in Chapter 2, "Division and the Art of Being Fair." These rules will help your child see what factors go into numbers without calculating.

AT THE KITCHEN TABLE

FACTOR FIGURES

For this game, you need an opponent, 2 dice, paper and pencil, masking tape, and small items to be used as counters: pennies, small candies, paper clips, and so on. Both players will use the same counters. Use the masking tape on one die to cover sides showing 1, 2, and 6. Write the numbers 7, 8, and 9 in the place of 1, 2, and 6. Now you are ready to play.

Before playing, roll both dice. Put the numbers together to make a two-digit number: 4 and 8 could be 48 or 84. If you choose 48, make a pile of 48 counters. When it is your turn, roll the die with tape on it. Suppose you get a 4. Try to arrange all 48 counters into even columns of 4. In mathematics, this is called an *array*. Start by placing 4 counters in a column. Then continue placing counters in even rows built out from the first column.

Here's how you would start your rectangle in this sample:

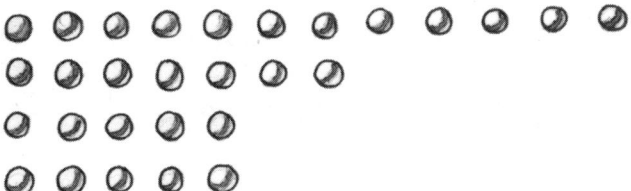

Keep filling in the rectangle until it is complete. In this case, you will get an array that is 4 counters high and exactly 12 counters wide, or 4 by 12. 4 and 12 are factors of 48. Write down 4 × 12 on your scorecard and give yourself one point.

IMAGINE THAT!

The number 37, which cannot be wholly divided by any number (except by 1 and itself), has the property that it will wholly divide the following numbers: 111, 222, 333, 444, 555, 666, 777, 888, 999.

Now it is your opponent's turn. Push the 48 counters back together. If your opponent rolls a 7, then he or she would place 7 of the 48 counters in a column and then place counters in rows built out from that first column. This is how the counters would look:

The counters do not fit evenly. Therefore, 7 is not a factor of 48. So, you have one point, and your opponent has none.

Keep playing until you think you have all the factors of 48. Then roll both dice again to start over with a new two-digit number.

On Your Own

Skinny and Plump

For this activity, you will need a pile of small items, such as buttons, pennies, or paper clips. Choose a prime number (one that has only 2 factors). Try 7, for example. Create an array (see At the Kitchen Table: Factor Figures) using the small items. For 7, you will find that you can make only one kind of array: a long, skinny rectangle.

Next choose a *composite number* with an *even number* of factors greater than two, such as 24. Experiment with making arrays. You will find that you can make many different arrays of many different dimensions. But remember, an array must always have an even number of columns and rows to show the factors of a number.

Finally, try a *composite number* with an *odd number* of factors, such as 36. Again, make as many different arrays as you can. For 37, make a square with 6 rows and 6 columns. Only composite numbers with an odd number of factors can make an array in the shape of a square. That's why they are called *square numbers*.

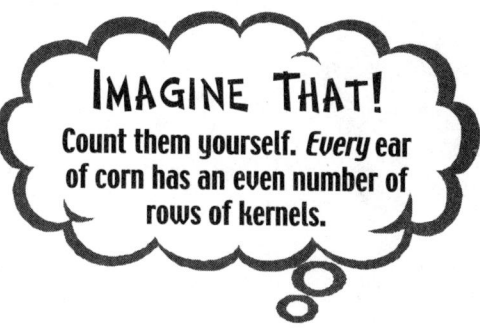

IMAGINE THAT!
Count them yourself. *Every* ear of corn has an even number of rows of kernels.

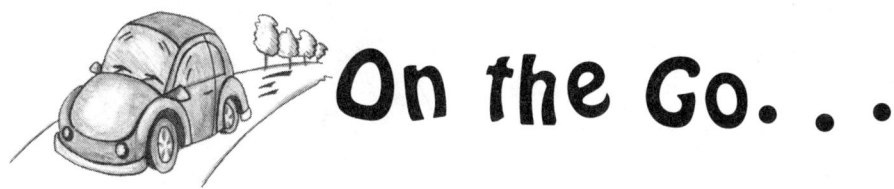

On the Go...

FACTOR FIND

Here is a game to help your child with factor knowledge. Give your child a number. Start easy and keep the first number below 100. Have your child tell you all the factors of that number. At first, let your child recite the factors in their pairs. For example, the factors of 12 are 1 and 12, 2 and 6, 3 and 4. As your child gains skill, have him or her present the factors in order from lowest to highest. For example, the factors of 10 are 1, 2, 5, 10.

Work with your child and go through the counting numbers to make sure that no factors were missed. Ask, "Does 1 go into this number?" The answer to this question will always be yes, because every number has 1 and itself as factors. So, list 1 and the number itself. "Does 2 go into this number? Is the number even?" Continue asking about each counting number until you come to a pair of factors that are very close to each other or that are one and the same—for example, the factors 4 and 6 for the number 24 and the factors 4 and 4 for 16. This means you've probably reached the end of the factors.

Congratulate your child and start with another number.

THE GREATEST COMMON FACTOR OF ALL

Parents' Corner

When you are first faced with the idea of a greatest common factor (GCF) in your child's studies, you will probably have to shake your head and hope that you will be able to retrieve the memory. But don't despair. In short, the GCF is the largest factor that will divide two separate numbers evenly. For example, the factors of 24 are 1, 2, 3, 4, 6, 8, 12, and 24. Factors of 28 are 1, 2, 4, 7, 14, and 28. The largest factor that's the same for both 24 and 28 is 4. So, 4 is the greatest common factor.

You need to help your child extend the ability of listing factors of numbers to determine the greatest common factor of two numbers. Being able to determine the GCF will help your child find like denominators so that he or she can add or subtract fractions easily.

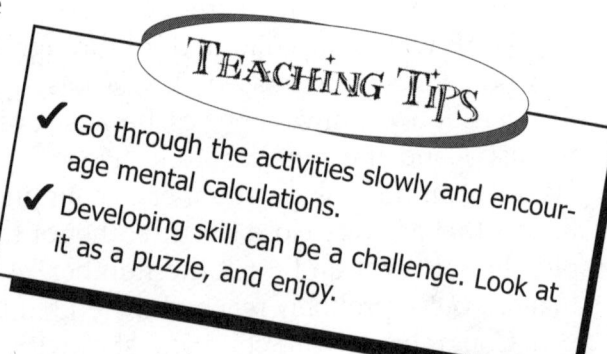

Teaching Tips

✓ Go through the activities slowly and encourage mental calculations.
✓ Developing skill can be a challenge. Look at it as a puzzle, and enjoy.

AT THE KITCHEN TABLE

GCF Concentration

Get a parent to play this fun game with you. You will need a score sheet, scratch paper, and a stack of index cards or paper with the numbers 1 through 50 written on them, one number on each card. Mix the cards thoroughly and lay them face down.

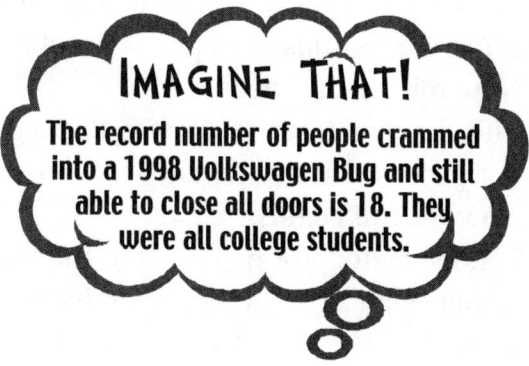

IMAGINE THAT!
The record number of people crammed into a 1998 Volkswagen Bug and still able to close all doors is 18. They were all college students.

When it is your turn, flip over any two cards. List the factors of each of those two numbers on scratch paper. Then identify their Greatest Common Factor. When you are sure that you have all the factors, clear your throat and make your Greatest Common Factor announcement. The GCF of the two numbers is your score. For example, if you choose 36 and 48, the factors would be as follows:

36: 1, 2, 3, 4, 6, 9, 12, 18, 36
48: 1, 2, 3, 4, 6, 8, 12, 16, 24, 48
Announce: "The GCF of 36 and 48 is 12" and record 12 as your score.

Each time it is your turn, add the new GCF to your score. Keep playing until someone scores 50 points.

On Your Own

Factor Fiction

To play this game, you need a pair of dice and a piece of paper. Roll the dice and write down the two numbers.

Let's say you roll a 4 and a 6. You can write 46 or 64 on the piece of paper—it's up to you! For this example, we'll choose 64.

Now roll the dice again.

This time we'll say that you rolled a 3 and a 2. You choose 32. Write it on the piece of paper, right next to the 64.

Now the fun begins.

List the factors for 64 below it. Remember, mathematicians list factors in order from smallest to largest. Then list the factors for 32 right next to the list of 64 factors.

64	32
1	1
2	2
4	4
8	8
16	16
32	32
64	

> **Imagine That!**
> The first millennium (A.D. 1–1000) consisted of 365,250 days. The second millennium (A.D. 1001–2000) consisted of 365,237 days. The current millennium (A.D. 2001–3000) will consist of 365,242 days. Leap years make all the difference.

Scan the list for factors the numbers have in common.

Circle the largest factor that they share. That is the greatest common factor. You found it!

Roll the dice again for two more numbers to factor.

On the Go...

GREATEST TRAVEL FACTOR

As you travel—whether by car, train, bicycle, or even foot—stay on the lookout for 2-digit numbers. Call out any two 2-digit numbers that you see. Have your child recite the common factors and end with the statement, "The greatest common factor of these two numbers is X!" Here is an example:

18 and 32: The common factors of these numbers are 1 and 2.
The greatest common factor of these numbers is 2.

Chapter 2

Division

In This Chapter

- Division and the Art of Being Fair
- The Power of Ten—in Reverse
- Daddy, Mommy, Sister, Brother . . . and More

DIVISION AND THE ART OF BEING FAIR

Parents' Corner

Most children understand the concept of fair share: "One for you, one for me." In fact, many kids invest great amounts of time and effort making sure they're dividing things "fairly." Use this intuitive understanding of division. Ask your child such questions as, "If we had four grapes, and you and I wanted to share them fairly, how many would we each get?" Divison is simply breaking apart a quantity into equal shares. Start small and build on these beginning concepts as your child learns about dividends and divisors.

Division is the inverse (or opposite) of multiplication. In mathematics, *inverse* means that one operation "undoes" the other. Just as multiplication was repeated addition, division is repeated subtraction. To divide 40 by 5, you can subtract 5 from 40 again and again until you reach 0. You will subtract it 8 times, so $40 \div 5 = 8$. Because multiplication and division are opposites, if your child knows the multiplication facts and can multiply all kinds of numbers, division will be much easier.

When working with your child, remember to use the division vocabulary:

> The *dividend* is the number being divided.
> The *divisor* is the number you are dividing by.
> The *quotient* is the answer in a division problem.
> 42 *(dividend)* ÷ 6 *(divisor)* = 7 *(quotient)*

Teaching Tips

✓ Emphasize the ways that multiplication and division are connected. Use counters to show that multiplication and division are opposite operations.

✓ Use correct terminology whenever possible: Dividend ÷ Divisor = Quotient.

AT THE KITCHEN TABLE

GAZINTA

In multiplication, you need to know your facts, or at least how to figure them out quickly. These facts will help you with division, too. In addition to using multiplication facts to find answers to division problems, there are also divisibility rules. These ruleswill let you look at a number and tell what "gazinta" (goes into) that number.

÷ 2   A number is divisible by 2 if it is even. This makes sense—if a number is even, it can be split into two equal groups. Which of the following are divisible by 2?

12   34   6   102

÷ 3   A number is divisible by 3 if, when you add all the digits together, you can divide the answer by three evenly. It's a great trick! Take 98 as an example. 9 + 8 = 17, and the sum of 1 and 7 is 8. Because 8 is not divisible by 3, 98 is not divisible by 3. But what about 96? 9 + 6 = 15, and 1 + 5 = 6. 6 is divisible by 3, and so is 96! Which of the following are divisible by 3?

27   34   42   105

÷ 4   A number is divisible by 4 if the last two digits are divisible by 4. The last two digits of a number are in the tens and ones places, so the last two digits would never be higher than 99. All even hundreds are divisible by 4, as are thousands, ten thousands, hundred thousands, millions, and so on. Look at 248. 248 is the same as 200 + 48. You know that 200 (100 + 100) is divisible by 4, so check 48. 48 divided by 4 is 24, so 248 is divisible by 4. Which of the following are divisible by 4?

336   254   572   188

÷ 5   A number is divisible by 5 if it ends in a 5 or a 0. This is an old rule, but it's one that works. Prove that it's true: count by 5s and see what kind of numbers you get! Which of the following are divisible by 5?

115   360   274   85

÷ 6   A number is divisible by 6 if it is even (or divisible by 2) and if you can add the digits together, divide by 3, and end up with no remainder, or leftovers. This rule blends together the divisibility rules for 2 and 3. Look at number 18. Because 18 is even, it is divisible by 2. The sum of the digits in 18 (1 and 8) is 9, and 9 is divisible by 3, so 18 is divisible by 3. Since 18 is divisible by 2 and 3, 18 is divisible by 6. This rule also makes sense if you think about it like this: $2 \times 3 = 6$. Which of the following are divisible by 6?

24   48   52   66

÷ 8   A number is divisible by 8 if the last three digits (the ones, tens, and hundreds place) are divisible by 8. This is a cousin to the 4 rule. Why do you have to check the last three digits? Well, 8 does *not* go into 100 evenly, as 4 does, but 8 *does* go into 1,000 evenly. One hundred 8s give you 800 (you need 200 more to reach 1,000). Twenty 8s give you 160 (160 + 800 = 960. You need 40 more to reach 1,000.). $5 \times 8 = 40$. 800 + 160 + 40 = 1,000. So 8 "gazinta" every thousand, every ten thousand, every hundred thousand, and every million. Look at the last three digits of 3,464. Does 8 go into 464 evenly? $8 \times 50 = 400$.

8 × 8 = 64. 400 + 64 = 464, so 8 "gazinta" 464. Therefore 8 also "gazinta" 3,464. Which of the following are divisible by 8?

6,832   10,042   1,156   3,264

÷ 9   A number is divisible by 9 if you can add the digits together and divide by 9 without getting a remainder, or leftover numbrs. In 918, 9 + 1 + 8 = 18, 1 + 8 = 9. 9 is divisible by 9, and so is 918. Which of the following are divisible by 9?

243   323   468   522

÷ 10   A number is divisible by 10 if it ends in a 0. That sounds easy enough! Which of the following are divisible by 10?

440   600   665   32,701

Divisibility Number Cruncher

Remember the t-chart from multiplication? This number crunching machine will also help you practice and solve problems with division. Find a partner (a parent, for example). Think of a division rule, such as ÷ 3—but don't tell your opponent! Draw a t-chart on a piece of paper. (Remember: A t-chart is a tool that mathematicians use to organize information.) Have your opponent give you numbers for the IN column. Then tell him or her what the OUT column is after you apply the rule. If your opponent says 12, you would apply ÷ 3 and end up with 4.

IN	OUT
12	4
21	7

Let your opponent keep putting in numbers until he or she has figured out the rule. If your opponent gives a number that is not divisible by 3, you can say that the number does not divide evenly.

Once your opponent guesses the rule, switch roles. Now *you* have to figure out what the function rule is. Have fun!

Variation

Here is another way to play this game. Fill in the IN and OUT information, but leave the IN slot empty for one and leave the OUT slot empty for another. Like this:

IN	OUT
32	4
80	10
16	?
?	5

Can your opponent guess the division rule of this function machine? Your opponent will have to multiply to figure out the missing information.

On Your Own

Daily Division

Division is a part of your daily life. For just one day, keep a tally sheet of how many times you find yourself in a situation when you need to divide. You might find yourself sharing food, toys, and time—dividing your food, your toys, and your time among different people. You'll be surprised at how much you use this operation every day!

On the Go...

TRAVELING COMPUTATION

This is a good game to play when you're driving in the car. Give your child a multiplication fact—such as 4 × 6 = 24. Your child's job is to recite the division facts that go with this fact family. In this case, 24 ÷ 6 = 4 *and* 24 ÷ 4 = 6.

Try this with many different multiplication and division facts.

JUST HOW FAR HAVE WE GONE?

Mileage provides a great real-life opportunity for children to practice estimating and dividing. Many drivers want to know about how many miles their cars get per gallon of gas. The next time you fill your tank, have your child look at the pump to note how many gallons went into your car. Next, have your child check the tripometer (the odometer that can be set back to 0 each time the car is filled with gas) and note the number it shows. For example, your child might note:

Gallons: 12.2   Miles driven: 313

Have your child estimate, by using division, about how many miles you've been able to drive per gallon of gas.

313 miles divided by 12.2

= about 300 (miles rounded) miles divided by 12 (gallons rounded)

Your child may estimate that the car got at least 20 miles to the gallon, since 20 × 12 = 240. The difference between 313 and 240 is 73. Since 12 × 6 = 72, the car got 20 miles plus 6 miles, or about 26 miles to the gallon.

If you ever take a long road trip, have your child check the gas mileage and see if you get better or worse mileage than when you drive short distances around town.

THE POWER OF TEN—IN REVERSE

Parents' Corner

Most children find it easy to multiply and divide by powers of ten (tens, hundreds, thousands, and so on). They are taught to count the 0s at the end of the number that they are multiplying by and *add* that many 0s or to count the number of 0s they are dividing by and take *off* that many. For example:

12 × 200   12 × 2 is 24. There are two 0s in 200, so the answer is 2,400

480 ÷ 60   48 ÷ 6 is 8. One 0 in 60 matches one 0 in 480, so the answer is 8.

Teaching Tips

✓ Make sure your child understands what happens when a number is divided by a multiple of 10.

✓ Discuss what "computating" means. If you divide 200 cupcakes evenly among 10 classrooms, each classroom will receive 20. After it is divided evenly, the 200 cupcakes become 10 smaller groups of 20: 200 ÷ 10 = 20.

✓ Encourage your child to use what he or she knows. If 48 ÷ 6 = 8, then 480 ÷ 6 = 80. Using division facts and dividing by the powers of ten will help your child.

This memory trick is good—but even if your child learns it, he or she may not know if the answer is correct. You must help your child understand what is happening.

When you multiply by powers of 10, numbers get very big very fast. When you divide by powers of 10, numbers get very *small* very fast. And just as in multiplication, being able to mentally calculate or estimate an answer will help your child with division problems that use many numbers. Take a look at this:

20 ÷ 10 = 2   200 ÷ 10 = 20   2,000 ÷ 10 = 200

See the pattern? In mathematics, as numbers add 0s, they grow 10 times, 100 times, 1,000 times, and so on. Division is the inverse, or opposite, of multiplication. This means that as numbers lose 0s, they decrease by 10 times, by 100 times, by 1,000, and so on.

AT THE KITCHEN TABLE

DIVISION CONCENTRATION

For this game, you will need a pile of blank cards on which you will write division problems and answers, a big table where you can spread out everything, and at least one opponent.

Write each of the following on one card, with problems and answers on separate cards:

2,000 ÷ 100	20	3,400 ÷ 10	340
920 ÷ 10	92	1,000 ÷ 1,000	1
720 ÷ 80	9	81,000 ÷ 900	90
450 ÷ 90	5	64 ÷ 8	8
3,000 ÷ 6	500	3,500 ÷ 70	50

As you get better at division, add more problems and answers to the pile.

Shuffle the cards and lay them on the table face down. When it is your turn, flip over two cards. If the two cards you choose show the problem and its correct answer, the pair is yours, and you get another turn. If the two cards do not match, turn them back over, but try not to forget where they are. The secret to concentration is memory. When it is your turn again, use what you know to match the problems and solutions. In the end, the person with the most matches wins. Good luck, and may your memory get a good workout!

On the Go...

ZEROS, ZEROS EVERYWHERE!

This is a great game to play while walking, making dinner, or driving in the car. It involves trading division challenges. Challenge your child to a series of division questions. Try a few of these, for example, and make some up, too!

   300 ÷ 10 (30)   4,200 ÷ 100 (42)   600 ÷ 100 (6)   50 ÷ 10 (5)

Take turns trying to stump one another with tricky division by powers of ten!

On Your Own

IT'S ALL IN THE HEAD

Try these problems, but *beware*—no writing allowed! You can do it all in your head.

1. 240 ÷ 6 = _____
2. 240 ÷ 60 = _____
3. 2,400 ÷ 60 = _____

4. 1,500 ÷ 3 = _____
5. 1,500 ÷ 30 = _____
6. 1,500 ÷ 300 = _____

7. 7,200 ÷ 9 = _____
8. 7,200 ÷ 90 = _____
9. 7,200 ÷ 900 = _____

10. 560 ÷ 80 = _____
11. 5,600 ÷ 80 = _____
12. 56,000 ÷ 80 = _____

Congratulations! You are becoming a great divider!

DADDY, MOMMY, SISTER, BROTHER... AND MORE

Parents' Corner

When your child learns long division, he or she has reached a huge milestone in elementary school. Some children see division as a challenge, whereas others fear its mere mention. In either case, many children find long division confusing, and mistakes will be made. The good news is that your child can make sense of long division and understand the process, instead of only learning the steps to follow when dividing.

People use many different algorithms when dividing. (An *algorithm* is a step-by-step process.) In the United States, most are taught the Daddy, Mommy, Sister, Brother method: **D**ivide, **M**ultiply, **S**ubtract, **B**ring down. In this example, you find a number that divides into part of the divisor. Multiply that number by the dividend. Subtract the product from the part of the divisor and bring down the next number in the divisor. And then continue the process.

```
        8
    42)3,400
    - 336
        40
```

Teaching Tips

✓ Be patient with your child. This is difficult stuff.
✓ Practice in short spurts, one or two problems at a time—not 10 or 15 problems at once.
✓ Use what you know about multiplication to help with division.

If that algorithm confuses your child, try another—multiplication in reverse—that makes division easier for those who already understand multiplication. This one is described in detail in the At the Kitchen Table activity in this lesson.

AT THE KITCHEN TABLE

Daddy, Mommy, Sister, Brother

Many people find it easier to remember the steps of long division with the help of the algorithm **D**addy (**D**ivide), **M**ommy (**M**ultiply), **S**ister (**S**ubtract), **B**rother (**B**ring down). Look at the problem 728 ÷ 23 to see how this works:

```
      31 R 15
  23)728
    - 69
      38
    - 23
      15
```

Divide. 23 can go into 72 three times.
Multiply. 3 × 23 = 69.
Subtract to get 3. **B**ring down the 8.
Divide. **M**ultiply 1 × 23.
Subtract. Since 15 is smaller than 23, this is leftover, or the remainder.

You made it! That's long division! Now try a few more problems following the same steps to see if you really get it.

1. 867 ÷ 13 = _____

2. 529 ÷ 37 = _____

3. 1,376 ÷ 24 = _____

4. 9,542 ÷ 12 = _____

Just for Fun!

Customer: How much is that duck?
Shopkeeper: Ten dollars.
Customer: Okay, could you please send me the bill?
Shopkeeper: I'm sorry, but you have to take the whole bird.

Chapter 2

MULTIPLICATION IN REVERSE

Here's a different way to look at division. Think of everything you know about multiplication. Multiplication will come in very handy, especially if you are good at multiplying by 10, 100, 1,000, and so on.

Even though this method takes longer than using the Daddy, Mommy, Sister, Brother method, you will always get the right answer if you use it. The more you practice, the better you will get at estimating how many of the divisor to put into the dividend.

Here are the steps. Look at the long division problem on page 45 and follow along. The list below matches up with the numbered circles on page 45.

1. Do one thousand 56s fit into 8,764? No, 56,000 is way too big! Do one hundred 56s fit? 56 × 100 = 5,600. Yes. Write 5,600 beneath 8,764. Then, next to the problem, keep track of the number of 56s you used (100).

2. Subtract 8,764 − 5,600. Can you fit another one hundred 56s in 3,164? No. Can you fit ten 56s in 3,164? Yes! Write 560, and record the number of 56s you used (10).

3. Subtract 3,164 − 560. Can you fit another ten? How about twenty? Figure out what 20 × 56 is by doubling 560 (10 × 56) or by figuring out 2 × 56. 20 × 56 = 1,120. Write 1,120 and record the number of 56s you used (20).

4. Subtract. Another twenty 56s will fit, you say? OK. Write 1,120 and record the number of 56s (20).

5. Subtract. 364 is not big enough to fit ten 56s in, but try 5. Just cut 10 × 56 in half to get 280. Don't forget to record the number of 56s used (5).

6. Subtract. How many 56s can you squish into 84? Only one? OK, but don't forget to record and subtract.

7. 28 is smaller than the divisor, so it is your remainder.

8. To find the answer to the problem, add up the number of 56s you used, and write that number at the top of the equation with the remainder.
100 + 10 + 20 + 20 + 5 + 1 = 156, with 28 leftover.

```
         156 R 28    ⑧
    56)8,764
      - 5,600   100  ①
        3,164
        - 560   10   ②
        2,604
      - 1,120   20   ③
        1,484
      - 1,120   20   ④
          364
         -280   5    ⑤
           84
         - 56   1    ⑥
           28        ⑦
```

Use multiplication in reverse to figure out the following problems:

1. 867 ÷ 13 = _____

2. 529 ÷ 37 = _____

3. 1,376 ÷ 24 = _____

4. 9,542 ÷ 12 = _____

On the Go...

ESTIMATE AND DIVIDE

When doing long division—or any kind of division, for that matter—it's important to know if an answer seems reasonable. Your child needs to make sure that all of his or her effort has resulted in the right answer. This game will help your child practice estimating answers to the nearest 10, 100, 1,000, and so on.

While running errands or doing dishes, give your child a division problem for estimation. For example:

452 ÷ 8

Your child does not need pencil and paper to figure this out. All that's required are some estimation skills. If necessary, prompt with questions like, How many 8s go into 452? Encourage your child to use what he or she already knows: 4×8 is 32, so 40×8 is 320. That's a little low. $5 \times 8 = 40$, and $6 \times 8 = 48$, so $50 \times 8 = 400$, and $60 \times 8 = 480$. The number 452 is between 400 and 480. Your child can give the very impressive answer, "It is between 50 and 60."

Try a few more problems for estimation, and then give your child a rest.

On Your Own

FOREIGN MATH

Different cultures have different ways to compute in mathematics. If you know kids who went to school in a different country, ask them how they multiply and divide. See if you can make sense of how they learned to divide, if it isn't the same method that your teacher used. If you don't know anyone from another country, do some research on the Internet or in an encyclopedia. Do you think methods of division from other countries are any easier than what you have learned today?

Fractions

In This Chapter

- Fraction, Fraction, What Is a Fraction?
- There's a Whole in This Fraction
- Where's the One (or the Two, for That Matter)?
- Improper Fractions: Not a Matter of Manners
- Where, Oh Where Is That Half?
- What's It Worth to You?
- Equivalents: Different but the Same

FRACTION, FRACTION, WHAT IS A FRACTION?

Parents' Corner

Whole numbers made perfect sense when it came to adding, subtracting, multiplying, and dividing—even when some of the calculations were difficult. With fractions, you must look at the parts in addition to the whole. This can cause frustration for both you and your child. As your child acquires new knowledge about fractions, try to make this knowledge as sensible as possible. Show your child how the numbers make sense so he or she will understand fractions.

Fractions are used to describe parts of a whole or parts of a group. For example, if Great Aunt Margaret cooks one of her delicious apple pies, and you cut the pie into eight *equal* pieces, you have just made fractions out of the whole apple pie. You cut the pie into eighths. The pieces are parts of a whole.

> **Teaching Tips**
> ✓ Review with your child the meanings of denominator and numerator.
> ✓ To explain a part-to-whole situation with a fraction, start with the phrase "out of": "Two out of three glasses on the table are empty." Then shift to fraction-speak: "Two-thirds of the glasses are empty."

A fraction consists of two numbers. The bottom number, or the *denominator*, tells the number of equal parts in the whole. The top number, or *numerator*, tells how many pieces you are paying attention to. If someone gave you $\frac{1}{3}$ of a candy bar, you know that the candy bar is cut into three equal pieces and that you got one of those pieces.

Here's a little help for remembering which is the top and which is the bottom number:

The **d**enominator is **d**own (the number on the bottom).

The numerator is the other number (the one on the top).

Fractions are simply another way to show division. The bar that separates the numerator from the denominator means to divide. $\frac{1}{2}$ means 1 whole divided into 2 equal parts. Each part is $\frac{1}{2}$ of the whole. $\frac{1}{3}$ could be 1 delicious chocolate cake divided by 3 hungry people. Each person gets $\frac{1}{3}$ of the cake.

AT THE KITCHEN TABLE

FRACTION ILLUSTRATION

For this game, you will need a pile of blank cards (small pieces of paper or index cards), a pencil, and someone to play with. Write a different fraction on each card:

$\frac{1}{2}$ $\quad$ $\frac{1}{3}$ $\quad$ $\frac{2}{3}$ $\quad$ $\frac{1}{4}$ $\quad$ $\frac{2}{4}$ $\quad$ $\frac{3}{4}$ $\quad$ $\frac{1}{5}$

$\frac{2}{5}$ $\quad$ $\frac{3}{5}$ $\quad$ $\frac{4}{5}$ $\quad$ $\frac{1}{6}$ $\quad$ $\frac{2}{6}$ $\quad$ $\frac{3}{6}$ $\quad$ $\frac{4}{6}$

$\frac{5}{6}$ $\quad$ $\frac{1}{8}$ $\quad$ $\frac{2}{8}$ $\quad$ $\frac{3}{8}$ $\quad$ $\frac{4}{8}$ $\quad$ $\frac{5}{8}$ $\quad$ $\frac{6}{8}$

$\frac{7}{8}$ $\quad$ $\frac{2}{2}$ $\quad$ $\frac{3}{3}$ $\quad$ $\frac{4}{4}$ $\quad$ $\frac{5}{5}$ $\quad$ $\frac{6}{6}$ $\quad$ $\frac{8}{8}$

Shuffle the cards, and put them in a stack, face down.

There are two versions to this game. For the first version, you will also need a pile of pennies. For the second version, you will need some scratch paper and pencils.

Version 1

When it is your turn, take a card from the stack and read your fraction. Let's say you picked $\frac{4}{5}$. Take 5 pennies and set them up so that 4 are showing heads. Say out loud: "4 out of 5 pennies are heads. $\frac{4}{5}$ of the pennies are heads."

Now it's your partner's turn. Suppose your partner turns over $\frac{7}{8}$. Your partner should

> **Just for Fun!**
>
> **Question:** If there are ten cats in a boat and one jumps out, how many are left?
> **Answer:** None. They were all copycats!

set up his or her pennies and say, "7 out of 8 pennies are heads. $\frac{7}{8}$ of the pennies are heads."

Keep playing until you run out of cards. Beware of the fractions that have the same number for the numerator and the denominator. $\frac{6}{6}$ is an interesting fraction. What will you do in that case?

Version 2

Set this version up in the same way as the first, but use paper and pencil instead of pennies. Suppose you pick $\frac{2}{3}$. Draw a rectangle and divide it into three equal sections. Shade two of the sections. Say out loud, "2 out of the 3 pieces of the rectangle are shaded. $\frac{2}{3}$ of the rectangle is shaded." Bravo!

On Your Own

Sandwiches to Go

For this activity, you will need paper and scissors.

The fraction $\frac{3}{4}$ represents 3 divided by 4. Let's see if that works. Tommy has 3 yummy sandwiches in his lunch box (he is very hungry today). As he sits down to begin lunch, his friends Susan, Natalie, and Nicole drop by. Tommy offers to share his sandwiches. If he wants to "fair share" (give the same amount to each of his three friends and himself), how much sandwich will each person get?

To help Tommy figure this out, cut 3 sandwiches (or squares) from your paper. Now cut each sandwich into fourths, and divide them equally among Tommy and his three friends. How many pieces does each person get?

On the Go...

Fraction Challenge

While you and your family are on the go, use fraction exercises to help strengthen your child's skills. Here are some examples:

Describe the cars parked in your dentist's parking lot. 3 out of 5 cars are white, so $\frac{3}{5}$ of the cars are white.

1 out of 2 pieces of your sandwich are left in your lunch box, so $\frac{1}{2}$ of your sandwich is still in your lunch box.

After some practice, add a little more challenge! This is a good way to pass the time when you have to wait—in line for a movie or at the bank to deposit checks. You go first. Describe the line ahead of you as a fraction: "2 out of 5 people wear glasses. $\frac{2}{5}$ of the people wear glasses."

Now your child has to use a different fraction to describe the line ahead in a different way—perhaps "1 out of 5 people in line wear a hat. $\frac{1}{5}$ of the people wear hats."

Keep playing until you reach the front of the line. Remember, the denominator will change as the line in front of you gets shorter.

THERE'S A WHOLE IN THIS FRACTION

Parents' Corner

You can use fractions in two ways to describe part-to-whole relationships. First, you can use them to describe the equal pieces of a single thing that has been cut up. For example, "Michael ate $\frac{2}{3}$ of the pizza" or "Daniel ate $\frac{3}{4}$ of the pizza." The pizza is a single whole item. Second, you can use fractions to describe parts of a group of many things that have been divided. For example, a fourth grade class has 30 children in it, and 19 of the children ride the bus home. $\frac{19}{30}$ of the class rides the bus. The whole here is made up of a group of 30 children.

Teaching Tips

- ✓ Keep talking about the importance of the whole as you enjoy the following activities.
- ✓ Use food examples whenever you can. Even a child who thinks the number 3 cannot be split in half will find a way to divide three cookies between two people.

This second possibility gets tricky when the problem involves a fraction of a group and we want to figure out how many that fraction represents. Here is an example:

> A team has 12 swimmers on it. One-third of the team competes in the backstroke race. How many swimmers compete in backstroke?

You need to figure out what $\frac{1}{3}$ of 12 is. Without getting into multiplication of fractions yet, draw the 12 members of the team:

The fraction of backstrokers is 1 out of 3 groups of swimmers, or $\frac{1}{3}$, so help your child arrange the swimmers into three groups:

    🏊🏊🏊🏊   🏊🏊🏊🏊   🏊🏊🏊🏊

One of these groups will compete in the backstroke race, so 4 swimmers will swim backstroke.

At the Kitchen Table

Group Parts

This game requires you to find half of a number. Some of the numbers are odd. At first, you might think, "Half of an odd number? That's impossible," but you will find that it's not so difficult after all. To play, you will need some paper for keeping score, a pencil, a pile of counters (such as pennies, blocks, beans), a playing partner, and some blank index cards.

To begin, write each phrase below on a separate index card:

8 bicycles	3 cookies	10 gifts	9 apples
4 coats	1 banana	2 lollipops	5 sandwiches
6 basketballs	7 strawberries	11 crackers	12 crayons

Shuffle the cards, and put them face down in a pile. When it's your turn, choose a card and read it to your partner: "6 basketballs." For this example, you would put out six counters to show the six basketballs. Take half of the basketball counters. How many did you take? Three. Write a 3 under your name on the score sheet.

Next, your partner gets a turn (maybe your grandmother is visiting for the week). Let's say that Grandma chooses "3 cookies." She puts out 3 counters and then attempts to take back half. How many should she take? If Grandma takes 2 cookies, is that half? What if she takes 1 cookie? Is that half? How would you solve this dilemma? If you say that Grandma should take 1 cookie and another $\frac{1}{2}$ of a cookie, you are correct. Grandma should take $1\frac{1}{2}$ cookies in all. So she would put $1\frac{1}{2}$ under her name on the score sheet. Now it is your turn.

The winner is the first person to reach 20. If you run out of cards, shuffle the used cards and make a new pile, or think of new cards that you could add to your deck. If you don't know how to add fractions yet, ask for help. And don't worry, we'll learn about adding fractions in the next chapter!

Half Truths

When working with fractions, keep in mind what the whole is, whether you're talking about pizza or fifth graders. Fractions describe parts of things you're looking at—but will $\frac{1}{2}$ always be the same $\frac{1}{2}$? This is a big question—and it has two answers. The first: Yes, $\frac{1}{2}$ will always be half of whatever you started with, or whatever the whole is. When you compare halves of two different wholes, however, they may look very different. This is a great conversation game to have with your family. Start by asking, "What is bigger—half of a mouse or half of an elephant?" After everyone stops giggling, someone will tell you, half an elephant is bigger than half a mouse. Ask why. Isn't $\frac{1}{2} = \frac{1}{2}$? Here are some other half truths to try:

Which is smaller? Half of a piece of rice, or half of a potato?

If you were really hungry, what would you want? Half of a grape, or half of a pizza?

What takes longer? Half of a school day, or half of a Saturday?

Think of some other examples on your own. Then mix in some $\frac{1}{4}$ or $\frac{3}{4}$ truths.

On Your Own

Half a Paper

A half is not always the same as half. That's because not all wholes are the same size. Get a section of the newspaper and open it so the two-page spread lies across the table. Then get a small piece of notepaper. If you were to wrap a shoebox and could use $\frac{1}{2}$ of either piece of paper, would it matter which half you took? This may seem like a silly question, but remember that in each case, you would have half a piece of paper. Why is half of the newspaper different from half of the notepaper?

WHERE'S THE ONE (OR THE TWO, FOR THAT MATTER)?

Parents' Corner

In the activities presented so far in this chapter, your child has been exploring the meanings of *fraction*. Fractions are equal pieces of a whole. The *denominator* tells the number of pieces in the whole, and the *numerator* tells the number of pieces you are paying attention to. Fractions are always determined by the whole. They can be shown in many ways, including drawings or counters. In this lesson, you will help your child see that fractions can also describe whole numbers.

Fractions describe parts of things, but those parts can add up to a whole. If you drew a picture on $\frac{1}{2}$ of a piece of paper, your drawing would be on 1 out of the 2 equal parts of the paper. If you decided to draw on the other side of the paper, your picture would cover 2 out of the 2 parts of the paper, or the whole piece of paper. So $\frac{2}{2} = 1$.

> **TEACHING TIPS**
> - Demonstrate fractions with apples, oranges, or pieces of paper. Show, for example, that 4 halves can be put together to make 2 wholes.
> - Talk about the division in fractions. $\frac{6}{3}$ means 6 divided by 3. $\frac{8}{4}$ means 8 divided by 4.

As another example, suppose Sarah has two apples. She cuts both apples in half. How many halves does she have? Four halves, which can be written as $\frac{4}{2}$. This means that Sarah has pieces that are halves of apples, which is what the denominator tells you. According to the numerator, Sarah has 4 of those pieces. So 4 apple halves is the same as 2 whole apples.

Another way to look at this apple dilemma is as a division:

$\frac{4}{2}$ means 4 divided by 2, which is 2.

Fractions in Disguise

Learning to understand fractions means traveling a long road, but you are moving swiftly along that road! This game will help you search for whole numbers disguised as fractions. At the same time, it will help you learn to recognize numbers greater than 1. To play, you will need a pad of paper, a pencil, and a partner.

Ask your partner to give you a fraction that is a whole number. Let's say your partner says $\frac{6}{2}$. Your task is to draw $\frac{6}{2}$ and determine how many wholes there are in $\frac{6}{2}$.

The denominator shows you that the whole is made of 2 parts, or halves. The numerator shows you how many halves you need: 6. Draw 6 halves. Then put your halves together to make wholes. You should have 3 wholes. Say out loud, "$\frac{6}{2}$ = 3 wholes." Well done! Now have your partner give you another fraction.

Fraction Concentration

To play this game, you will need a pile of blank pieces of paper to make cards, a pencil, scratch paper on which to draw your fractions, and a partner.

Copy each of these fractions and numbers on the blank pieces of paper. Put one fraction or number on each piece of paper:

$\frac{9}{3}$ $\frac{8}{4}$ $\frac{14}{2}$ $\frac{10}{2}$ $\frac{10}{10}$

$\frac{12}{3}$ $\frac{18}{3}$ $\frac{40}{5}$ $\frac{18}{2}$ $\frac{30}{3}$

1   2   3   4   5

6   7   8   9   10

Shuffle the cards and lay them out face down in four rows of five cards each. When it is your turn, flip over two cards. Does each card have the same number written on it? If so, prove it. For example, if you choose 2 and $\frac{8}{4}$ and you say they are equal, explain why that is true. Drawing the fraction might help. Dividing 8 by 4 might also help. If you match two cards, and you are able to prove it, keep the pair of cards.

What if you flipped over $\frac{10}{2}$ and 3. Are they the same? Draw 10 halves. Then put the halves together to make wholes and count how many you have. This is not a match. Put the cards back where you found them, face down, but remember where they are in case you pick a match for one of them when your turn comes round again.

When all matches have been made, total up the number of matched cards you have to see who wins.

IMAGINE THAT!
During a 24-hour period, the average human will breathe 23,040 times, exercise 7 million brain cells, and speak 4,800 words.

Challenge

After you get really good, make this game a little harder by adding these cards to the stack you already have:

$\frac{15}{5}$	$\frac{16}{8}$	$\frac{21}{3}$	$\frac{25}{5}$	$\frac{15}{15}$
$\frac{8}{2}$	$\frac{12}{2}$	$\frac{16}{2}$	$\frac{27}{3}$	$\frac{100}{10}$
1	2	3	4	5
6	7	8	9	10

In this version, you might run into some very interesting situations. What if you turn over a 4 and a 4? Are they the same number? Of course! You made a match. But what if you turn over $\frac{40}{5}$ and $\frac{16}{2}$? Are they the same number? Don't forget: You'll have to prove it before you get your point.

On Your Own

CRACK THE CODE

You know a lot about fractions. You are developing fraction sense. Now it's time to jiggle your sense a little to see if you can figure out some missing denominators and numerators.

You have learned that whole numbers can hide in fractions. In this activity, you need to crack the code and find the missing numerator or denominator to make the equation true. Sketch out the pieces if it helps you to figure it out.

1. $\frac{?}{8} = 2$ 2. $\frac{?}{4} = 1$ 3. $\frac{?}{6} = 3$

4. $\frac{4}{?} = 2$ 5. $\frac{9}{?} = 3$ 6. $\frac{12}{?} = 2$

Make up a few yourself and have one of your parents try it!

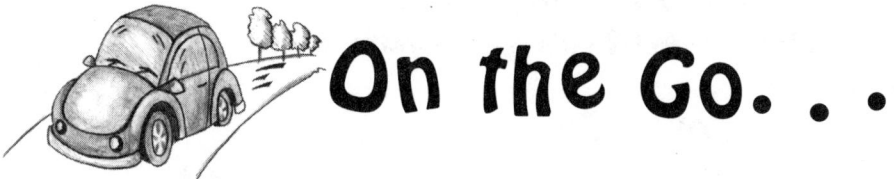

On the Go...

STOP TO THINK ABOUT IT

Kids get bored riding around town in the back seat. Here's something they can do to add some pizzazz to those errand trips. At every stoplight or stop sign, tell your child a denominator and then ask the following questions or add a few of your own:

> Fourths.
> What is 1 in fourths? ($\frac{4}{4}$)
> What is 2 in fourths? ($\frac{8}{4}$)

If questions like these are too difficult, ask:

> How many fourths do you need to make two wholes? (8)
> How many fourths do you need to make three wholes? (12)

Change the denominator at every stop.

IMPROPER FRACTIONS: NOT A MATTER OF MANNERS

Parents' Corner

As your child becomes comfortable with whole numbers written as fractions, throw in this curve: What about fractions that are larger than 1 but that do not divide evenly into whole numbers? What about something like $\frac{3}{2}$, for example? Or $\frac{8}{3}$? These fractions, called improper fractions, get their name not because they are without manners or taste, but because their numerator is bigger than their denominator. This pattern should alert you that whenever you see a numerator that is larger than a denominator, you automatically know that the fraction represents at least a whole number. These fractions are like little puzzles.

> **TEACHING TIPS**
> ✓ Use the activities to help your child understand improper fractions and mixed numbers and how they are the same.
> ✓ Whenever possible, express a number both as an improper fraction and as a mixed number.

Fractions that are improper when written as a single fraction can also be written (or *expressed*) as *mixed numbers*. In this form, they are made up of a whole number and a fraction. For example,

$$\frac{3}{2} = 1\frac{1}{2} \qquad \frac{8}{3} = 2\frac{2}{3}$$

The interesting thing about mixed numbers and improper fractions is that they can change form at a moment's notice. $1\frac{1}{3} = \frac{4}{3}$, or vice versa. $\frac{8}{5} = 1\frac{3}{5}$, or vice versa. How does this work? Remember the whole: $1\frac{1}{3}$ has $\frac{1}{3}$ and 1 whole in it. How many thirds are in 1 whole? $\frac{3}{3} = 1$ whole. Now count your thirds. $\frac{3}{3}$ for the whole, plus $\frac{1}{3} = \frac{4}{3}$ all together.

AT THE KITCHEN TABLE

QUICK DRAW

For this game, you will need scratch paper, pencils, a stack of blank index cards that you'll make into fraction cards, and another player. To make the cards, write one improper fraction or mixed number on each card:

$1\frac{3}{4}$ $8\frac{1}{2}$ $4\frac{2}{8}$ $6\frac{1}{4}$ $3\frac{2}{5}$ $1\frac{1}{2}$ $3\frac{2}{3}$ $7\frac{1}{3}$

$\frac{7}{2}$ $\frac{5}{3}$ $\frac{8}{5}$ $\frac{9}{2}$ $\frac{6}{4}$ $\frac{10}{3}$ $\frac{11}{6}$ $\frac{15}{7}$

Shuffle the cards and put them face down in a pile. Each player should be ready with a pencil and a piece of paper.

To begin, one player turns over a card and reads it out loud: "$\frac{6}{4}$." Both players have 10 seconds to think about the fraction before the player that chose the card says, "Go!" As quick as your fraction fingers will allow, draw $\frac{6}{4}$. One way to do this would be to draw a square. Divide it into fourths and shade $\frac{6}{4}$. Is it possible? No! So draw another square, divide it into fourths, and shade fourths until you reach $\frac{6}{4}$. When you are done, put your pencil down and wait for your opponent to finish. When you've both finished drawing, the person with the first pencil down explains why his or her picture shows $\frac{6}{4}$. The opponent must listen, question, and agree (if the explanation is correct). Then the person with the second pencil down explains his or her picture. If a picture is wrong, work together to fix the picture.

Start the second round and draw a new fraction.

On Your Own

Mixed Sandwiches

Here's a problem for you to solve. Use paper and scissors to figure out the answers. In this case, cutting is better than drawing, because you can move the pieces around.

> Suppose Christine and Kenneth made some sandwiches for their study party. To make things interesting, they cut each sandwich into fourths and made a pile of fourths on a plate. Their friends were happy to sample a variety of sandwiches and ate while they studied. After everyone went home, Christine and Kenneth cleaned up. They had some sandwiches left. To be exact, they had seven fourths left. How many whole sandwiches were left over? How many fourths?

Use your paper and scissors to figure this out. Think about how many fourths would make a whole sandwich. Make a whole sandwich using the pieces. Can you make another whole sandwich? When you have made all of the whole sandwiches that you can from seven fourths, how many do you have? How many fourths are left over?

Imagine That!
End to end, the number of Crayola crayons made in a year would circle the globe $4\frac{1}{2}$ times. You could draw a lot of pictures with that!

How Many?

This game is good to play over and over again for a couple minutes at a time. When you are out pulling weeds (or whatever errands you are running that day), ask your child, "How many eighths in 2?" Be sure to start with a whole number to warm up. Then make the questions a little harder. After your child tells you $\frac{16}{8}$ as the answer to the first question, ask how many more eighths are in $2\frac{3}{8}$. Three more eighths adds up to $\frac{19}{8}$. Next, your child takes a turn asking you a question—maybe "How many fourths in 3? How many fourths in $3\frac{1}{4}$?"

Keep playing until you have pulled a pile of weeds (or finished folding the laundry, or whatever you were doing).

WHERE, OH WHERE IS THAT HALF?

Parents' Corner

As your child learns more about fractions, he or she will learn to recognize some of them regardless of form, no matter how they are disguised. A child who understands the relationship of numerator to denominator will easily spot equivalent fractions of the basics—$\frac{1}{2}, \frac{1}{3}, \frac{1}{4}$, and even $\frac{3}{4}$. For example, in $\frac{1}{2}$, the numerator (1) is half of the denominator (2). Think of another fraction that is half. How about $\frac{5}{10}$? Is the numerator (5) half of the denominator (10)? It is. $\frac{5}{10}$ is therefore an equivalent fraction to $\frac{1}{2}$.

You can help prove this by drawing pictures or using food (the great explainer) to prove equivalent fractions. If you have two oranges, slice the first one in half. Show that two halves make one whole. Explain that 1 is half of 2 (1 + 1 = 2). Cut the other orange into ten slices. Separate the slices into two groups of 5. Then show that 5 is half of 10.

The following activities can provide good practice for understanding this relationship for each of the basic fractions. Don't worry about other fractions for now. Your child will have plenty of time to practice them later.

Teaching Tips

✓ Use $\frac{1}{2}$ the first time you do the following activities. As your child becomes comfortable with fractions equivalent to $\frac{1}{2}$, explore and practice with other basic fractions.

✓ Repeat activities until they become easy.

AT THE KITCHEN TABLE

THE HALF SHUFFLE

For this game, you'll need a stack of blank index cards, some scratch paper, pencils, and at least one other player. Write one number and word on each card:

5—numerator	6—denominator	6—numerator
7—numerator	8—numerator	8—denominator
9—numerator	10—numerator	10—denominator
2—numerator	2—denominator	3—numerator
4—numerator	4—denominator	12—denominator

Shuffle the cards and put them face down in a pile. When it is your turn, take a card from the deck. Let's say you choose "8—denominator." Figure out what fraction equivalent to $\frac{1}{2}$ has 8 as a denominator. Draw a picture to prove it—or draw the picture as you try to figure out the answer. Then use this picture to prove your answer.

How much is shaded? $\frac{4}{8}$? Yes! Show the other players your picture, and say "$\frac{1}{2} = \frac{4}{8}$." After the other players give you the thumbs up, another player may take a turn.

What would happen if you picked "9—numerator" on your next turn? You would then have to figure out what the denominator would be to make a fraction equivalent to $\frac{1}{2}$ with 9 as the numerator. So $\frac{9}{?} = \frac{1}{2}$. You know that the numerator of fractions equivalent to $\frac{1}{2}$ is always half of the denominator, so the denominator would be 18. Put your findings on paper by drawing a picture of $\frac{9}{18}$. Say out loud, "$\frac{1}{2} = \frac{9}{18}$." (Show your picture at the same time.) Again, you get a unanimous thumbs up!

The Quarter Shuffle

Play this game the same way as the Half Shuffle, but make cards with the following information to find equivalents for $\frac{1}{4}$:

2—numerator	3—numerator	4—numerator
4—denominator	5—numerator	6—numerator
7—numerator	8—numerator	8—denominator
9—numerator	10—numerator	11—numerator
12—denominator	16—denominator	20—denominator

The Third Shuffle

Since you love this game so much, you can also play it looking for equivalents for $\frac{1}{3}$. Here are the cards for the game:

2—numerator	3—numerator	3—denominator
4—numerator	5—numerator	6—numerator
6—denominator	7—numerator	8—numerator
9—numerator	9—denominator	10—numerator
12—denominator	15—denominator	18—denominator

On the Go...

Name That Half

This game will help your child understand $\frac{1}{2}$ wherever it appears. As you scurry through your daily routines, name a fraction type for your child. For example, you might say, "Fourths!" Your child can say one of two things: "Impossible" (if it is impossible to name $\frac{1}{2}$ in fourths) or "I can name that half." In the latter case, encourage your child to picture a favorite candy bar divided into 4 equal pieces. Can he or she take half of those pieces? If so, then half would be $\frac{2}{4}$, and that is what your child should say.

Try another example. What if you say, "Sevenths"? Your child should picture that yummy candy bar neatly divided into 7 pieces. Can it be divided in half? It is impossible to divide the sevenths evenly without cutting them up again, so your child should say, "Impossible."

Challenge: Name That Fraction

As your child becomes comfortable with $\frac{1}{2}$ and the relationship between the numerator and the denominator, try other basic fractions. Play the game using $\frac{1}{4}$. Discuss the relationship between numerator and denominator. Instead of the numerator being half of the denominator, as in $\frac{1}{2}$, the denominator in $\frac{1}{4}$ is 4 times the numerator. So you might say "Eighths," and your child would say, "$\frac{2}{8}$."

For thirds, the denominator is 3 times the numerator. So if you say "Ninths," your child should say "$\frac{3}{9}$." This is more of a challenge, but it will help your child develop a sense of how fractions work.

On Your Own

Fraction Draw

Compare $\frac{1}{2}$ and $\frac{5}{10}$. Draw a rectangle and draw a line down the middle so the rectangle is cut in half. Shade $\frac{1}{2}$ of the rectangle. Directly below the first rectangle, draw another whole rectangle exactly the same size. Divide that rectangle into 10 equal pieces. Shade 5 of the pieces. Compare the two rectangles. They both have the same amount ($\frac{1}{2}$) shaded.

Do this same activity for the following fractions.

1. $\frac{1}{2} = \frac{4}{8}$

2. $\frac{1}{2} = \frac{3}{6}$

3. $\frac{1}{2} = \frac{2}{4}$

> **Just for Fun!**
>
> How many seconds are there in a year?
> 12. The 2nd of January, 2nd of February, etc.

WHAT'S IT WORTH TO YOU?

Parents' Corner

In their quest to understand, children learn to know fractions as they know whole numbers. They know that 5 is bigger than 2 but smaller than 10. They can put whole numbers in order and find their approximate locations on a number line. Although ordering fractions is not as easy as that, your child can learn to do this, too.

How big is $\frac{3}{4}$? Is it bigger than $\frac{1}{3}$? What fractions are between $\frac{1}{5}$ and $\frac{1}{2}$? If you can put fractions in order, you will have a better sense of how much you are talking about. As an example, Mr. Riley owns a hardware store. He is organizing a new shipment of nails and wants to put them on the wall by length. He wants to put the smallest nails at one end and the longest nails at the other end. He could open all of the packages and compare the nails, but open packages don't sell very well. He looks at the packages and sees that the size is clearly marked. He picks up two packages: One says $\frac{3}{4}$ inch, and the other says $\frac{1}{2}$ inch. Which is bigger? Which is smaller?

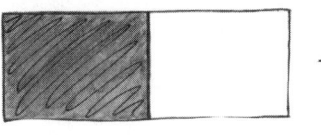

With a good understanding of fractions, you and your child will be able to sort out Mr. Riley's nails.

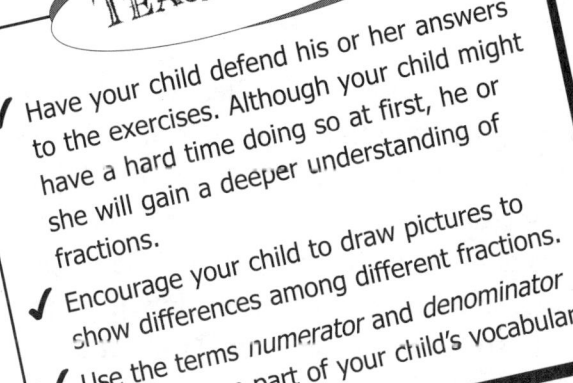

Teaching Tips

- ✓ Have your child defend his or her answers to the exercises. Although your child might have a hard time doing so at first, he or she will gain a deeper understanding of fractions.
- ✓ Encourage your child to draw pictures to show differences among different fractions.
- ✓ Use the terms *numerator* and *denominator* to make them part of your child's vocabulary.

AT THE KITCHEN TABLE

ALLIGATOR MUNCH

For this game, you will need scratch paper, pencils, another player, and blank index cards. Draw a wide-open alligator mouth on one of the index cards. Don't forget the teeth! Set this card aside.

Write a different fraction on each index card:

$\frac{1}{2}$ $\frac{2}{2}$ $\frac{1}{3}$ $\frac{2}{3}$ $\frac{3}{3}$ $\frac{1}{4}$ $\frac{2}{4}$ $\frac{3}{4}$

$\frac{4}{4}$ $\frac{1}{5}$ $\frac{2}{5}$ $\frac{3}{5}$ $\frac{4}{5}$ $\frac{5}{5}$ $\frac{1}{6}$ $\frac{2}{6}$

$\frac{3}{6}$ $\frac{4}{6}$ $\frac{5}{6}$ $\frac{6}{6}$ $\frac{1}{8}$ $\frac{2}{8}$ $\frac{3}{8}$ $\frac{4}{8}$

$\frac{5}{8}$ $\frac{6}{8}$ $\frac{7}{8}$ $\frac{8}{8}$ $\frac{1}{12}$ $\frac{2}{12}$ $\frac{3}{12}$ $\frac{4}{12}$

$\frac{5}{12}$ $\frac{6}{12}$ $\frac{7}{12}$ $\frac{8}{12}$ $\frac{9}{12}$ $\frac{10}{12}$ $\frac{11}{12}$ $\frac{12}{12}$

Shuffle the fraction cards, and place them face down in a pile. When it is your turn, flip over the top two cards. Use the alligator mouth as a greater than or less than sign to show which fraction is larger. Prove that you are right by drawing a picture on scratch paper. For example, if you turn over $\frac{2}{3}$ and $\frac{1}{8}$, draw each of these fractions and show which is larger. The alligator will always eat the larger amount. Look at your drawings and put the alligator mouth so that it opens to the larger fraction: If you drew "$\frac{2}{3}$" and "$\frac{1}{8}$", you would show "$\frac{2}{3} > \frac{1}{8}$." Say out loud, "$\frac{2}{3}$ is greater than $\frac{1}{8}$," and show your picture. Then give your partner a turn.

On the Go...

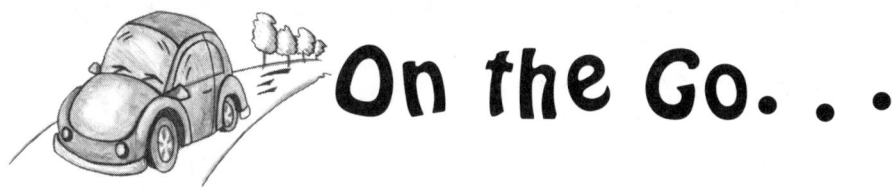

Are You First?

This game requires ordering fractions. This does not mean driving up to a fast-fraction place and ordering a $\frac{5}{8}$, $\frac{2}{3}$, and $\frac{1}{4}$ to go—and hold the catsup, please. This kind of order means to put fractions in order by size.

As you drive along or wait for an appointment, give your child two fractions. At first, give fractions with the same denominators: "$\frac{2}{5}$ and $\frac{4}{5}$," for example. Your child's job is to put the fractions in order from smallest to largest and prove that he or she is right. For example, your child might say, "$\frac{2}{5}$ is the smaller fraction, and $\frac{4}{5}$ is the bigger fraction. $\frac{2}{5}$ is only 2 out of 5 equal parts, but $\frac{4}{5}$ is 4 out of 5 equal parts." Show that you are impressed, and call out two more fractions: "$\frac{6}{10}$ and $\frac{1}{10}$."

As your child improves, increase the number of fractions to three or four. Then try fractions with different denominators: $\frac{3}{4}$ and $\frac{1}{3}$, $\frac{1}{2}$ and $\frac{1}{8}$, or $\frac{6}{7}$ and $\frac{9}{10}$. (*Note:* Ordering fractions with mixed denominators might be tough to do on the go. You may want to give your child paper and pencil to sketch out the fractions.)

On Your Own

Let Me Eat Cake

You have always learned that bigger means more. In fractions, that rule does not fit. When it comes to fractions, in fact, the bigger the denominator, the smaller the fraction.

Try this: On two separate pieces of paper, draw two identical pictures of your favorite cake. Cut one picture into 8 equal pieces. You get $\frac{1}{8}$ of the cake. (Is your mouth watering?) Now cut the other into 14 equal pieces. You get $\frac{1}{14}$ of the cake. Since fourteen is greater than eight, are you happy with the $\frac{1}{14}$? Is bigger always better?

EQUIVALENTS: DIFFERENT BUT THE SAME

Parents' Corner

Equivalent means equal in value. When you work with whole numbers, you could say that you packed five things to read on your summer vacation—or you could say that you have three magazines, one mystery novel, and one book about North American history. Either way, you have five things to read. You have said the same thing in different—but equivalent—ways.

Likewise, equivalent fractions name the same amount in different ways. They are fractions that are equal in value yet that look different. For example, $\frac{1}{4}$ is the same as $\frac{2}{8}$, which is the same as $\frac{5}{20}$. All fractions have many disguises. Understanding these disguises will help your child find equivalents. This, in turn, will help your child add or subtract fractions.

Here's an easy rule (or algorithm) to follow: "Whatever you do to the top, you need to do to the bottom." This is more like a fairness pledge than a rule since fractions are always about being fair and equal. For two fractions to be equivalent, the numerator and denominator must be treated the same. For example, to find an equivalent fraction to $\frac{4}{5}$, multiply both the top and the bottom by a number, let's say 3: $\frac{4}{5} \times \frac{3}{3}$. Because $\frac{3}{3}$ is the same as 1, you are not changing the value of $\frac{4}{5}$, since multiplying by 1 does not change the value of the number. $\frac{4}{5} \times 1 = \frac{4}{5} \times \frac{3}{3}$. Multiply the numerators and the denominators to find the new equivalent fraction: $\frac{4}{5} \times \frac{3}{3} = \frac{4 \times 3}{5 \times 3} = \frac{12}{15}$. $\frac{4}{5} = \frac{12}{15}$

> **TEACHING TIPS**
>
> ✓ Emphasize that equivalent fractions have the same value, even when they look different; and that multiplying by 1 does not change the value of the fraction.
>
> ✓ If your child is unsure about the fractional form of 1, go back to "Where's the One?" in this chapter for more practice.
>
> ✓ Ask your child to prove equivalents by drawing; this will tell you a lot about what he or she understands.

AT THE KITCHEN TABLE

SPINNING EQUIVALENTS

For this game, you will need two homemade spinners. Making these spinners is easier than making chocolate chip cookies—although cookies would go well with this game, too. To make the spinners, you need two pieces of paper, two paper clips, markers or crayons, a lid to a plastic bowl (not bigger than the paper), and a ruler.

Trace around the lid on both pieces of paper so you have a circle on each. Put a dot in the middle of each circle. Use your ruler to draw a line that cuts the circle in half, then in fourths, then in eighths. Repeat for the other circle.

Now, with your markers, write these fractions in the sections on one of the circles:

$$\frac{2}{3} \quad \frac{1}{3} \quad \frac{1}{2} \quad \frac{1}{4} \quad \frac{3}{4} \quad \frac{2}{5} \quad \frac{1}{6} \quad \frac{1}{8}$$

Don't throw this spinner away after you are done. It can also be used for "Spin and Compute" in chapter 4.

On the other circle, write these "ones"—one in each section:

$$\frac{2}{2} \quad \frac{3}{3} \quad \frac{4}{4} \quad \frac{5}{5} \quad \frac{6}{6} \quad \frac{7}{7} \quad \frac{8}{8} \quad \frac{10}{10}$$

Now it's time to play the game. When it is your turn, put a paper clip over the dot in the center of the fraction circle and stick the point of your pencil so it holds the paper clip on the dot. Flick the paper clip with your finger to spin it. Next spin the other spinner. Let's say you spin $\frac{2}{5}$ and $\frac{10}{10}$. Multiply the two together to find the equivalent.

$$\frac{2}{5} \times \frac{10}{10} = \frac{2 \times 10}{5 \times 10} = \frac{20}{50}$$

What an interesting fraction! Does it make sense that $\frac{20}{50} = \frac{2}{5}$? Why? Draw $\frac{2}{5}$. Try to picture $\frac{20}{50}$. Do these two fractions take up the same amount of space? If so, then you have found the equivalent. Now let your partner have a turn.

On the Go...

FRACTION WORKOUT

This activity will work the mental muscles your child needs to find equivalents. Have your child say a fraction to you—maybe $\frac{2}{3}$. You will say, "Multiply by $\frac{2}{2}$." Your child's job is to find the equivalent:

$$\frac{2}{3} \times \frac{2}{2} = \frac{2 \times 2}{3 \times 2} = \frac{4}{6}$$

After giving you the correct answer, have your child name another fraction, such as $\frac{3}{5}$. You could then say $\frac{3}{3}$. What is the equivalent?

$$\frac{3}{5} \times \frac{3}{3} = \frac{3 \times 3}{5 \times 3} = \frac{9}{15}$$

Try some more.

On Your Own

TWO (OR MORE) OF THE SAME

Look up equivalent in the dictionary. You will find that equivalent things have the same value. You may have a nickname or two for yourself, or your parents may call you a shortened version of your name (unless you are in trouble, and then they use all of your names in a rather loud way). These names are equivalent—they mean the same thing.

Equivalent fractions are also equal in value. They are the same fraction with a different name: $\frac{1}{2}$ can be $\frac{3}{6}$, but it can also be $\frac{10}{20}$ or even $\frac{50}{100}$! No matter what you call it, the fraction is half of the whole.

List ten equivalent fractions for each of the following:

1. $\frac{1}{2}$ 2. $\frac{1}{4}$ 3. $\frac{1}{3}$ 4. $\frac{3}{4}$

Working with Fractions

In This Chapter

- Add a Little, Subtract a Little
- Sensible Multiplication
- The Rules of Multiplication
- Division Made Easy
- The Rules of Division

ADD A LITTLE, SUBTRACT A LITTLE

Parents' Corner

Once your child understands basic fractions, adding and subtracting fractions will be easy. All prior knowledge of naming fractions, knowing the value of fractions, working with mixed numbers and improper fractions, and finding equivalents will come together to make adding and subtracting fractions a sensible task.

Let's say you wanted to add $\frac{1}{5} + \frac{2}{5}$. The denominators are the same, so you simply add the numerators: $\frac{1}{5} + \frac{2}{5} = \frac{3}{5}$. The same is true of subtraction with common denominators. $\frac{7}{8} - \frac{2}{8} = \frac{5}{8}$.

The extra work comes when you add or subtract fractions with unlike denominators: $\frac{2}{3} + \frac{1}{4}$. Many intermediate-age students make the common error of adding both numerators and denominators: $\frac{2}{3} + \frac{1}{4} = \frac{3}{7}$. This could just be a momentary freeze that can be thawed by a few strategic questions: "What did you say?" If not, suggest trying to make sense of the sizes of the fractions. Your child may see the mistake without even figuring out the problem exactly. If you have $\frac{2}{3}$ of a candy bar and your friend gives you $\frac{1}{4}$ of her candy bar, would you have less than $\frac{2}{3}$ of a candy bar? Explain to your child that $\frac{3}{7}$ is less than half of a whole candy bar, which is less than $\frac{2}{3}$.

> **TEACHING TIPS**
> - ✓ Play the games in steps. Stay at a step until your child fully understands what is taught.
> - ✓ After moving to a later step, sneak in a few problems or challenges from earlier steps.
> - ✓ Draw examples, or ask your child to draw to show how he or she got the answer.

You may want to break this lesson into four parts:

1. Adding/subtracting fractions with common denominators
2. Adding/subtracting fractions with unlike denominators
3. Adding/subtracting mixed numbers with common denominators
4. Adding/subtracting mixed numbers with unlike denominators

AT THE KITCHEN TABLE

BUILDING PROBLEMS

Building problems can be a good way to test your skills and keep you thinking. To do these, have your parent give you a starting number. Then your parent will give you a series of operations to end at another number. Here's a simple example:

Think of the number 17. Double it. Add 100. Subtract 50.

What's your answer? 84. How did you get there? Go back through the steps. You had 17, and you doubled it. $17 \times 2 = 34$. You added 100 to 34 to get 134. Then you subtracted 50 from 134. That left 84.

Common Denominator Building Problems

You can also build problems with fractions. Although this may sound difficult, you can do it if you start slowly. Use paper and pencil to draw pictures if you need to. Have your parent give you a series of tasks to do with fractions. Here's an example:

Think of the fraction $\frac{1}{8}$. Triple it. Add $\frac{2}{8}$ to it. Double your answer.

Did you come up with $\frac{10}{8}$? That's right. How did you get there? You started with $\frac{1}{8}$ and you tripled it: $\frac{1}{8} + \frac{1}{8} + \frac{1}{8} = \frac{3}{8}$. Remember to add only the numerators. The denominator stays the same. Then you added $\frac{2}{8}$ to $\frac{3}{8}$ to get $\frac{5}{8}$. For the finale, you doubled $\frac{5}{8}$, which gave you $\frac{5}{8} + \frac{5}{8} = \frac{10}{8}$. $\frac{10}{8}$ is an improper fraction. How many times does 8 go into 10? Once. How many are left over? 2. So the mixed number is $1\frac{2}{8}$, or $1\frac{1}{4}$.

IMPROPER FRACTION!

Here's another sample problem:

Think of the fraction $\frac{3}{4}$. Subtract $\frac{1}{4}$. Double your answer.

That was short and sweet. Did you get 1? How did you do it? You started with $\frac{3}{4}$ and subtracted $\frac{1}{4}$, which left you with $\frac{2}{4}$. Then you doubled your answer. Since you are good at finding $\frac{1}{2}$ wherever it hides, you knew that $\frac{2}{4} = \frac{1}{2}$, and if you double $\frac{1}{2}$ ($\frac{1}{2} + \frac{1}{2} = \frac{2}{2}$) you get 1. Yahoo!!

Here are some more to try. If you and your parents start to get tired or dizzy, rest a while. These problems can be pretty tough!

1. Think of the fraction $\frac{3}{4}$. Double it. Subtract $\frac{1}{4}$.
2. Think of the fraction $\frac{3}{8}$. Add $\frac{5}{8}$. Add $\frac{7}{8}$. Subtract $\frac{1}{8}$.
3. Think of the fraction $\frac{2}{3}$. Add $\frac{2}{3}$. Double it.

Mixed Numbers With Like Denominators Building Problems

Now try these:

4. Think of the number 8. Add $\frac{1}{4}$ to it. Add $1\frac{1}{4}$ to your answer.
5. Think of the fraction $\frac{1}{4}$. Triple it. Double your answer. Now add $1\frac{3}{4}$.
6. Think of the fraction $\frac{5}{8}$. Double it. Add 2 to your answer. Subtract $1\frac{1}{8}$ from your answer.

Just for Fun!

Andrae: What do you call two owls that jump out of a plane?

Megan: Pair-a-hoots!

SPIN AND COMPUTE

For this game, spinner 1 is the same as the fraction spinner from the "Spin an Equivalent" game in chapter 3. If you haven't made that spinner already, go back and do it now. While you're at it, make two more. Label each spinner section with a fraction from the list on page 79:

Spinner 1 (already done)	Spinner 2	Spinner 3
$\frac{2}{3}$	$\frac{1}{5}$	$1\frac{1}{4}$
$\frac{1}{3}$	$\frac{4}{5}$	$2\frac{3}{8}$
$\frac{1}{2}$	$\frac{2}{6}$	$3\frac{1}{6}$
$\frac{1}{4}$	$\frac{5}{6}$	$4\frac{1}{2}$
$\frac{3}{4}$	$\frac{3}{5}$	$1\frac{2}{3}$
$\frac{2}{5}$	$\frac{3}{8}$	$2\frac{1}{3}$
$\frac{1}{6}$	$\frac{7}{8}$	0
$\frac{1}{8}$	$\frac{1}{3}$	1

You can play this game in various ways. Each way focuses on adding and subtracting fractions and mixed numbers that have unlike denominators. If you need help, use paper and pencil to draw the fractions. Your parent might also ask you to draw one or two of the equations to show how you thought through the answer. Don't worry: You're ready for this fraction challenge!

Fractions with Unlike Denominators

Use spinners 1 and 2. Put a paper clip over the middle of the spinner. Put your pencil in the paper clip, and spin the clip. Write the fraction where the clip lands ($\frac{1}{6}$, for example). Spin the second spinner and add that fraction to the first fraction. Say the second spinner lands on $\frac{1}{3}$. Write the equation you have created: $\frac{1}{6} + \frac{1}{3} = ?$.

To figure out the problem, search for a common denominator. What number do both 3 and 6 go into? Did you decide on sixths? Begin to compute the equivalents:

$\frac{1}{6} = \frac{1}{6}$ This one is free

$\frac{1}{3} = \frac{?}{6}$ $\frac{1}{3} \times \frac{2}{?} = \frac{2}{6}$ (be sure to multiply both the numerator and the denominator)

$\frac{1}{6} + \frac{2}{6} = \frac{3}{6}$

On your next turn, subtract the fractions that you spin. This might be a little tricky. You need to decide which fraction is bigger before you can subtract. Use what you know about the value and size of fractions. If you need to, ask your parent for help in figuring this out.

Mixed Numbers With Unlike Denominators

Now you can test your skill with mixed numbers that have unlike denominators. Play the same game, but this time use spinner 3 twice or use it once and use one of the other spinners once. You can do it!

When you're done, save the spinners for the game "Practice, and More Practice" in the sections on multiplication and division with fractions.

> **IMAGINE THAT!**
> The amount of time an average man spends shaving: 3,350 hours. The number of whiskers on the face of the average man: 30,000.

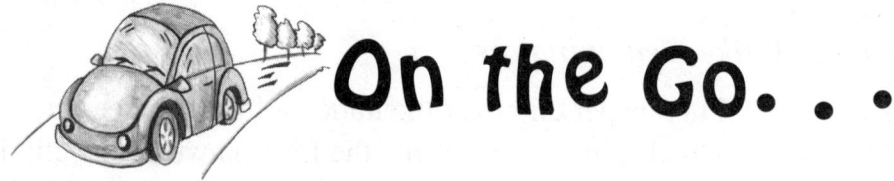

HEAD PROBLEMS

Head problems don't hurt—usually. Doing a head problem simply means that your child adds, subtracts, multiplies, or divides numbers mentally. Head problems have two, three, or four steps that build on each other. They are basically the same thing as Building Problems, except that your child does them mentally, without paper and pencil.

Start with simple problems, then make them more complicated as your child gets better at solving them. Here's an example:

Think of $\frac{1}{4}$. Double it. Add $\frac{2}{4}$. Subtract $\frac{1}{4}$. What do you end up with? ($\frac{3}{4}$)

On Your Own

COOKING WITH FRACTIONS

The best place to practice adding and subtracting fractions is in the kitchen. Cooks have to know about measurements because they often double and halve recipes. Help out whenever someone in your house is baking or cooking. Who knows? You might even get to lick the bowl!

Here's a recipe project for practice:

> Ms. Kelsie owns a bakery. She hired Bobby and Ray, two of her bakers, because of their ability to add and subtract fractions quickly. Ms. Kelsie likes to double and triple recipes, and when she is feeling really creative, she likes to combine recipes! Today she wants to double a chocolate chip cookie recipe. Here are some of the ingredients she needs for one batch of cookies:
>
> $\frac{3}{4}$ cup of butter
> $1\frac{1}{2}$ cups of brown sugar
> $1\frac{3}{4}$ cups of flour

What should Bobby and Ray do to double the recipe? See if you can help them. Look at the butter first. How much butter will they need altogether? Now double the other ingredients. Beware of improper fractions, and change them to mixed numbers.

> Now Ms. Kelsie is feeling a bit creative. She has decided to combine her zucchini bread recipe with the chocolate chip cookie recipe. (Zucchini chip cookies—they could be a new taste sensation!) Zucchini bread requires $\frac{2}{3}$ cups of butter, and chocolate chip cookies need $\frac{3}{4}$ cups of butter. Bobby and Ray know they cannot add these fractions yet.

Because the denominators are not the same, Bobby and Ray will need to find equivalents with common denominators for each of the fractions. The first question they ask is, What number can both 3 and 4 go into evenly?

Can you help Bobby and Ray? How much butter do they need? If zucchini bread calls for $1\frac{1}{2}$ cups of flour, how much flour will they need total? Remember that chocolate chip cookies need $1\frac{3}{4}$ cups of flour.

SENSIBLE MULTIPLICATION

Parents' Corner

Before learning the rules of multiplying fractions, first focus on how the process works.

Multiplying fractions is different from multiplying whole numbers. When you multiply fractions of less than 1, you get a number that is less than or equal to the numbers you multiplied.

$\frac{1}{3} \times \frac{3}{4} = \frac{3}{12}$, or $\frac{1}{4}$ $\frac{1}{4}$ is less than both $\frac{1}{3}$ and $\frac{3}{4}$.

When you multiply by a fraction, you take only part of it, or a fraction of it. With whole numbers, you want "4 groups of 6." With fractions, you want "$\frac{1}{3}$ of a group of $\frac{3}{4}$." That's pretty small. This strange way of thinking may seem upside down to your child, but with the help of this lesson, it should start making sense.

As an example, Tia wants to make necklaces. She has 3 feet of chain. She wants to use $\frac{1}{2}$ of the chain for one necklace. She needs to figure out $\frac{1}{2} \times 3$, or $\frac{1}{2}$ of 3. (The multiplication sign also means *of*.) This easy problem can be done in your head. $\frac{1}{2}$ of 3 is $\frac{3}{2}$, which is also $1\frac{1}{2}$. This also works when you need a fraction of a fraction. Say you need $\frac{1}{2}$ of $\frac{6}{8}$. Half of $\frac{6}{8}$ must be $\frac{3}{8}$.

A more complicated problem is $\frac{1}{3} \times \frac{3}{4}$. You want to find $\frac{1}{3}$ of $\frac{3}{4}$. To figure this out, draw a rectangle, and divide it into four pieces. Shade in three of the pieces. You want 1 of these 3 shaded pieces, or $\frac{1}{3}$ of them. Since there are three-fourths shaded, you get one of these fourths. $\frac{1}{3}$ of $\frac{3}{4}$ is $\frac{1}{4}$, or $\frac{1}{3} \times \frac{3}{4} = \frac{1}{4}$.

Teaching Tips

- ✓ Avoid showing your child the multiplying rule; that will come in the next lesson. This lesson is to help your child develop some understanding of how to multiply fractions.
- ✓ Draw problems to show what is happening.
- ✓ Discuss the logic behind each problem as you do the activities.

At the Kitchen Table

Half a Fraction

For this activity, draw pictures to prove that your multiplication is correct and that you understand why you are doing what you do. If the activity seems hard, keep practicing. Read the story about Ben and his sandwich. As you read the story, draw pictures to illustrate what happens. Then try the problems at the end.

Ben has a sandwich—his all-time favorite of bananas, peanut butter, and raisins. He has been thinking about this sandwich all morning. When the lunch bell finally rings, he snatches his lunch bag and finds a place to eat. He is unwrapping his delicious sandwich when his sister Ellice sits down next to him. She looks a little sad. She has forgotten her lunch, and her stomach is completely empty. Ben offers her half of his sandwich. She happily thanks him and munches down her half. Ben's mouth is watering. Just as he is about to take a bite of his half, his friend Andre sits down. Andre tells Ben how he lost his lunch. He has looked everywhere, but it seems to be gone. Ben kindly offers Andre $\frac{1}{2}$ of his $\frac{1}{2}$ sandwich. Andre accepts. Ben now has $\frac{1}{4}$ of his sandwich left.

What if another hungry friend comes by? Then Ben will be faced with $\frac{1}{2}$ of $\frac{1}{4}$. How much will Ben have left? $\frac{1}{8}$ is right—only a biteful or two.

Try drawing a few of these problems on a piece of paper. Draw the second fraction first, and then cut it in half:

1. $\frac{1}{2}$ of $\frac{8}{10}$ **2.** $\frac{1}{2} \times \frac{1}{4}$ **3.** $\frac{1}{2} \times \frac{3}{4}$ **4.** $\frac{1}{2}$ of $\frac{1}{3}$

Some of these require more brain power than others. Remember that fractions are the fairest folks around: do to the top what you do to the bottom.

If you feel a little too challenged, take a break and relax. When you've given your brain a rest, try to draw a few of these:

5. $\frac{1}{4} \times \frac{1}{3}$ **6.** $\frac{1}{4} \times \frac{1}{5}$ **7.** $\frac{1}{4} \times \frac{4}{5}$ **8.** $\frac{1}{4} \times \frac{1}{2}$
9. $\frac{1}{3} \times \frac{3}{4}$ **10.** $\frac{1}{3} \times \frac{1}{3}$ **11.** $\frac{1}{3} \times \frac{1}{2}$ **12.** $\frac{1}{3} \times \frac{1}{8}$

On the Go...

Why Bother?

Mathematicians often want to find the quickest ways to figure things out. If they can do a calculation mentally, why bother figuring it out on paper? This game is short and fast and will help develop your child's mental skills. You could try it anywhere—in the car, while taking a bath, or even when washing windows.

Start with halves. Ask your child to tell you half of each whole number up to twenty. Help your child to see the logic in what is happening.

$\frac{1}{2}$ of 1 ($\frac{1}{2}$) $\frac{1}{2}$ of 2 (1) $\frac{1}{2}$ of 3 ($1\frac{1}{2}$) $\frac{1}{2}$ of 4 (2)   ... $\frac{1}{2}$ of 20 (10)

Next try $\frac{1}{4}$. Instead of taking half, your child will take $\frac{1}{4}$ of the number.

$\frac{1}{4}$ of 1 ($\frac{1}{4}$) $\frac{1}{4}$ of 2 ($\frac{2}{4}$, or $\frac{1}{2}$) $\frac{1}{4}$ of 3 ($\frac{3}{4}$) $\frac{1}{4}$ of 4 (1)   ... $\frac{1}{4}$ of 20 (5)

If your child is really good at this, try $\frac{1}{3}$. This is more difficult.

$\frac{1}{3}$ of 1 ($\frac{1}{3}$) $\frac{1}{3}$ of 2 ($\frac{2}{3}$) $\frac{1}{3}$ of 3 (1) $\frac{1}{3}$ of 4 ($\frac{4}{3}$, or $1\frac{1}{3}$)

On Your Own

SHRINKING NUMBERS

You have probably noticed that multiplying fractions smaller than 1 is different from multiplying whole numbers. When you multiply whole numbers, the product is greater than or equal to the larger of the two numbers (unless you are multiplying by 0). $8 \times 2 = 16$. 16 is bigger than both 8 and 2. In multiplying fractions, the product gets smaller than the two numbers. $\frac{1}{3} \times \frac{1}{2} = \frac{1}{6}$. $\frac{1}{6}$ is smaller than both $\frac{1}{3}$ and $\frac{1}{2}$.

Let your brain ease into this new thinking. You will understand. Try these problems:

1. $\frac{1}{4} \times \frac{1}{2} =$ _____

2. $\frac{1}{3} \times \frac{1}{4} =$ _____

3. $\frac{1}{2} \times \frac{1}{2} =$ _____

> **Just for Fun!**
>
> **Question:** What two things can't you have for breakfast?
>
> **Answer:** Lunch and dinner.

THE RULES OF MULTIPLICATION

Parents' Corner

Multiplication is probably the easiest of the four basic operations to use with fractions. Now that your child is beginning to understand what happens in multiplying fractions, it is time to learn the multiplication rule for fractions. This rule will come in handy for problems that your child has trouble figuring out mentally or with a quick draw on paper. The rule, or algorithm, is simple: When multiplying fractions, multiply the numerator by the numerator and the denominator by the denominator.

$$\frac{3}{4} \times \frac{7}{8} = \frac{3 \times 7}{4 \times 8} = \frac{21}{32}$$

You simply multiply straight across, to get the answer.

The rule works the same for multiplying a fraction by a whole number.

$$\frac{5}{6} \times 3$$

This is a little different, but not much. A whole number is also a fraction. The denominator of a whole number as a fraction is 1, so $3 = \frac{3}{1}$. Let's try that problem again:

$$\frac{5}{6} \times 3 = \frac{5}{6} \times \frac{3}{1} = \frac{5 \times 3}{6 \times 1} = \frac{15}{6}, \text{ or } 2\frac{3}{6}, \text{ or } 2\frac{1}{2}$$

When your child multiplies mixed numbers, it is best to convert the mixed number to an improper fraction first. To figure out the improper fraction, simply multiply the denominator by the whole number and then add the numerator. So $2\frac{1}{3} \times 1\frac{1}{2}$ would become $\frac{7}{2} \times \frac{3}{2}$. Multiplying numerators and denominators results in $\frac{21}{6}$, or $3\frac{3}{6}$, or $3\frac{1}{2}$ (see chapter 3).

TEACHING TIPS

✓ Encourage your child to do these problems mentally whenever possible.

✓ Use the terms *denominator* and *numerator* frequently in your discussions.

AT THE KITCHEN TABLE

Practice, and Practice More

For this game, you will need spinners 1 and 2 from the "Spin and Compute" game in the "Add a Little, Subtract a Little" lesson of this chapter. You will also need a paper clip for the spinners, a pencil, scratch paper, and a playing partner.

When it is your turn, spin each spinner once. Record the two fractions, and put a multiplication sign between them to create a multiplication problem. Your problem might look like this: $\frac{3}{5} \times \frac{2}{3}$. You can do this in your head or on paper:

$$\frac{3 \times 2}{5 \times 3} = \frac{6}{15}, \text{ or } \frac{2}{5}$$

Record your number on a score sheet and continue playing until someone reaches 10.

Challenge Version

To play another version of this game, you could use spinners 1 and 3 from "Spin and Compute." Now you will create multiplication problems with mixed numbers. Spin each spinner once and see what you get.

Say you get $\frac{1}{3} \times 4\frac{1}{2}$. You first need to convert the mixed number to an improper fraction. Multiply the denominator by the whole number and then add the numerator to get $\frac{9}{2}$. Your new equation will look like this: $\frac{1}{3} \times \frac{9}{2}$. Ready to multiply?

$$\frac{1 \times 9}{3 \times 2} = \frac{9}{6} \text{ (improper fraction!), or } 1\frac{3}{6} \text{ (half!), or } 1\frac{1}{2}$$

For an even greater challenge, use only spinner 3, but spin it twice. Write down the two mixed fractions, and get ready to multiply. For $2\frac{1}{3} \times 1\frac{1}{4}$, convert the two mixed numbers to improper fractions and multiply:

$$\frac{7}{3} \times \frac{5}{4} = \frac{7 \times 5}{3 \times 4} = \frac{35}{12}, \text{ or } 2\frac{11}{12}$$

On the Go...

RULES, RULES, RULES

The more fraction multiplication your child does in the head, the more efficient he or she will become at mathematics. While running errands, running water, or running the mile, have your child try these problems:

$\frac{1}{3} \times \frac{1}{5}$ ($\frac{1}{15}$) $\frac{1}{6} \times \frac{1}{9}$ ($\frac{1}{54}$) $\frac{2}{3} \times \frac{1}{8}$ ($\frac{2}{24}$, or $\frac{1}{12}$) $\frac{1}{10} \times \frac{5}{6}$ ($\frac{5}{60}$, or $\frac{1}{12}$) $\frac{3}{8} \times \frac{1}{7}$ ($\frac{3}{56}$)

Think of more problems. Practice, practice, and practice some more to sharpen your child's mind and make him or her a fraction champion!

On Your Own

SIMPLY FRACTIONS

Mathematicians—and regular people, too—like to use fractions in their simplest form. Why say $\frac{5}{10}$ when you know you really mean $\frac{1}{2}$? Sometimes it is easy to see that the fraction is equivalent to $\frac{1}{2}$ or $\frac{1}{3}$ or $\frac{1}{4}$, but other times you need to compare the numerator and the denominator. Look at $\frac{8}{12}$. Can it be *simplified* (or *reduced*, or *put in lowest terms*)? Ask, "Does any single number go into both the numerator and denominator evenly?" Yes! 4 goes into the top, and 4 goes into the bottom.

$$\frac{8}{12} \div \frac{4}{4} = \frac{8 \div 4}{12 \div 4} = \frac{2}{3}$$

Therefore, $\frac{8}{12}$ and $\frac{2}{3}$ are equivalent fractions. $\frac{2}{3}$ is in the simplest form.

You can divide top and bottom by the same number and not change the value of the fraction because $\frac{4}{4}$ is the same as 1, and any number divided by 1 does not change it's value. Find the simplest form of these fractions:

1. $\frac{10}{20}$ _____   2. $\frac{12}{16}$ _____   3. $\frac{6}{8}$ _____   4. $\frac{16}{24}$ _____

DIVISION MADE EASY

Parents' Corner

Division with fractions can make many people shudder. Most people were taught to invert, or flip over, the second fraction and then multiply—but few have any idea what they are doing or why. Dividing fractions has an important link to your child's knowledge of whole numbers. It doesn't matter if you are dividing by 4 or by $\frac{1}{4}$: The point is to figure out how many of a number will fit into another—or how many of a number "gazinta" the other. For example, if you have 8 cookies and 4 children, how many cookies will each child get if you share the cookies evenly? 4 goes into 8 two times, so the children get 2 cookies each. The same idea applies to division with fractions.

A big difference between dividing by whole numbers and dividing by fractions of less than 1 is in the answer (or *quotient*). With whole numbers, the quotient is usually smaller than the number being divided (the *dividend*) because you are dividing a larger quantity into smaller pieces. With fractions, the quotient is greater than or equal to the dividend. You are dividing by chunks smaller than 1, so you will have more of them.

This reverse logic might confuse your child. Proof through objects your child can see and count—known as "manipulatives"—or through drawing is especially crucial in this lesson.

Teaching Tips

- ✓ Avoid showing your child the invert-and-multiply rule for now. Simply help your child understand the process of division with fractions.
- ✓ Discuss real-life division stories. If you have 3 pies each cut into fourths, how many people can you serve? ($3 \div \frac{1}{4} = 12$)
- ✓ Draw division problems or use manipulatives to improve your child's understanding.

At the Kitchen Table

Gazinta, Divide, and Draw

For this game, you need paper, pencils, a set of blank index cards for making equation cards, and another player. To make equation cards, write a different equation on each index card.

$\frac{1}{2} \div \frac{1}{4}$ $\qquad$ $\frac{5}{3} \div \frac{1}{3}$ $\qquad$ $\frac{7}{8} \div \frac{1}{4}$ $\qquad$ $\frac{4}{6} \div \frac{1}{3}$ $\qquad$ $\frac{3}{4} \div \frac{1}{2}$

$2\frac{1}{3} \div \frac{2}{3}$ $\qquad$ $\frac{6}{9} \div \frac{1}{3}$ $\qquad$ $\frac{1}{2} \div \frac{1}{8}$ $\qquad$ $\frac{8}{10} \div \frac{1}{5}$ $\qquad$ $\frac{3}{8} \div \frac{1}{8}$

Shuffle the equation cards and place them face down in a pile. When it is your turn, take an equation card. Your job is to read the equation, say it in the "gazinta" (goes into) way, draw the division to prove it, and tell the answer. Some of these are quick and easy, and some are more involved. Let's try a tricky one.

$\frac{7}{8} \div \frac{1}{4}$

Read the equation to the other players. Then say, "This problem is asking how many $\frac{1}{4}$s go into $\frac{7}{8}$." Draw a rectangle and divide it into eighths. Shade 7 of them.

Here's the tricky part. Divide the same rectangle into fourths, but let the fourth lines hang over the sides of the rectangle.

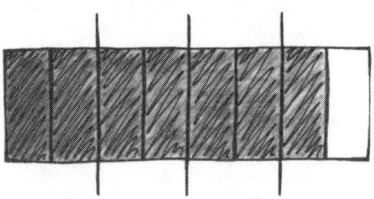

Count the $\frac{1}{4}$ pieces. There are three whole $\frac{1}{4}$ pieces, and there is $\frac{1}{2}$ of a $\frac{1}{4}$. So the answer is $3\frac{1}{2}$. $3\frac{1}{2}$ fourths fit into $\frac{7}{8}$.

Keep playing until you run out of cards. Then take a break.

> **IMAGINE THAT!**
> George Washington's inauguration speech was 183 words long and took 90 seconds to read.

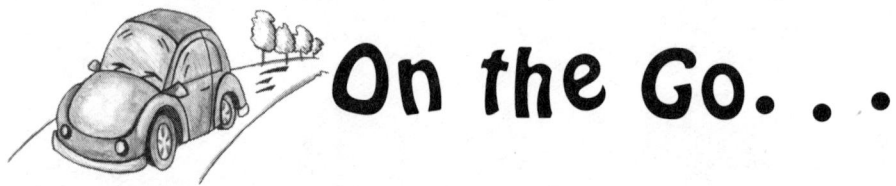

On the Go...

Why Bother? (Division Style)

Just as you can multiply some fractions in your head, you can divide some, too. All whole numbers divided by *unit fractions* (fractions with 1 in the numerator) can be done this way. Try a few with your child:

$3 \div \frac{1}{4} = 12$

A good way to help your child figure this out is to think about how many of the divisor (in this case, $\frac{1}{4}$) will fit into 1. Since four $\frac{1}{4}$s fit into 1, then $\frac{8}{4} = 2$, and $\frac{12}{4} = 3$. Try a few more:

$5 \div \frac{1}{2}$ (10) $10 \div \frac{1}{6}$ (60) $12 \div \frac{1}{3}$ (36) $7 \div \frac{1}{4}$ (28)

When your child can compute this kind of division problem quickly, try the next challenge. Ask your child to try some problems like this:

$5\frac{1}{2} \div \frac{1}{2}$ (11) $10\frac{1}{6} \div \frac{1}{6}$ (61) $12\frac{1}{3} \div \frac{1}{3}$ (37) $7\frac{1}{4} \div \frac{1}{4}$ (29)

Encourage your child to use what he or she has learned from dividing whole numbers to help solve mixed-number division problems.

$10\frac{3}{6} \div \frac{1}{6}$ (63) $12\frac{2}{3} \div \frac{1}{3}$ (38) $7\frac{3}{4} \div \frac{1}{4}$ (31)

On Your Own

Math Wiz

Division of fractions is not easy. And you're not only learning how to divide fractions, you are also learning how to divide some of them in your head—which takes even more brain power! This is an admirable skill, oh great mathematician. To keep your skill strong, here are a few to practice. Use what you know from the first problem to do the rest.

1. $2 \div \frac{1}{4} =$ _____ $2\frac{1}{2} \div \frac{1}{4} =$ _____ $2\frac{4}{8} \div \frac{1}{4} =$ _____

2. $\frac{4}{8} \div \frac{1}{4} =$ _____ $\frac{5}{8} \div \frac{1}{4} =$ _____ $\frac{3}{8} \div \frac{1}{4} =$ _____

Don't worry if you get stuck along the way. Each problem helps you think about solving the problem that comes after it. Run around the block. Get a drink of water. Let's try another set.

3. $\frac{6}{8} \div \frac{1}{8} =$ _____ $\frac{4}{8} \div \frac{1}{8} =$ _____ $\frac{1}{2} \div \frac{1}{8} =$ _____

4. $\frac{1}{4} \div \frac{1}{8} =$ _____ $\frac{3}{4} \div \frac{1}{8} =$ _____ $1\frac{1}{2} \div \frac{1}{8} =$ _____

Wow, there were a few big leaps in that set of problems! But you are a great mathematician, so have faith in yourself!

THE RULES OF DIVISION

Parents' Corner

"Ours is not to reason why: Just invert and multiply!" Remember that? Many successful adults still rely on this poem to see them through division of fractions. Although such algorithms are helpful shortcuts, they are, unfortunately, not *your child's* shortcuts. Someone else discovered the whys and hows of the shortcut and took away the understanding process from all who learn and use it. The algorithm depends on blind faith. But your child will know better, because he or she now understands the connection between the division of whole numbers and the division of fractions.

> **Teaching Tips**
> ✓ Practice using the algorithm frequently, but don't overdo it: two or three problems at a sitting are enough.
> ✓ Have your child estimate answers before figuring them out.

Division of fractions can be simple, as in the previous lesson ("Division Made Easy"), or it can be sticky, tricky, and long. Although all fraction division problems can be drawn, it is not always an easy task. For example,

$$\frac{5}{6} \div \frac{2}{3}$$

Draw a rectangle divided into 6 pieces and shade 5 of them. This gives you the dividend. Then divide the same rectangle into thirds and show $\frac{2}{3}$. Your divisor, $\frac{2}{3}$, would fit in once, with 1 section left—$\frac{1}{4}$ of the $\frac{2}{3}$ chunk. You might figure out that $\frac{2}{3}$ goes into $\frac{5}{6}$ $1\frac{1}{4}$ times.

Or you could use the algorithm invert and multiply.

$$\frac{5}{6} \div \frac{2}{3} = \frac{5}{6} \times \frac{3}{2} = \frac{5 \times 3}{6 \times 2} = \frac{15}{12} = 1\frac{3}{12} = 1\frac{1}{4}$$

Understanding what is happening is crucial to developing math sense. Recognizing when a problem is more involved, estimating the answer, and checking it with the algorithm are also valuable skills.

AT THE KITCHEN TABLE

FLIP AND MULTIPLY

One way to divide fractions is to invert the divisor (flip the fraction upside down) and then multiply the two numbers. To practice the invert-and-multiply algorithm, you have to practice the invert-and-multiply algorithm. Find a quiet place to work. Estimate your answers first. Write your estimate down on your paper. When you have finished a set of three, explain each one to your parent, including your estimate and your reason for your estimate, work, and answer. After you do three, go clean out your closet. Then come back and do three more.

1. $\frac{3}{4} \div \frac{1}{3}$ Estimate. _____   Work.

2. $1\frac{2}{3} \div \frac{4}{5}$ Estimate. _____   Work.

3. $\frac{7}{8} \div \frac{1}{3}$ Estimate. _____   Work

96 Chapter 4

4. $\frac{8}{9} \div \frac{3}{4}$ Estimate. _____   Work.

5. $1\frac{4}{5} \div \frac{1}{8}$ Estimate. _____   Work.

6. $\frac{6}{11} \div \frac{2}{5}$ Estimate. _____   Work.

7. $2\frac{2}{3} \div \frac{3}{8}$ Estimate. _____   Work.

8. $\frac{9}{10} \div \frac{2}{3}$ Estimate. _____   Work.

9. $\frac{6}{8} \div \frac{1}{7}$ Estimate. _____   Work.

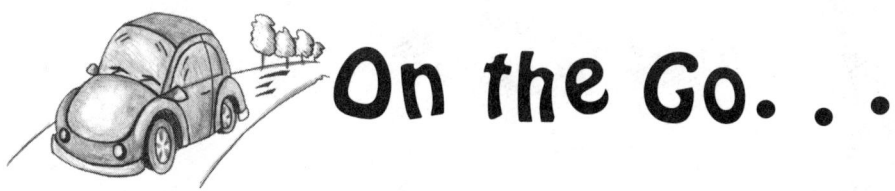

On the Go...

WHY BOTHER? (DIVISION STYLE 2)

You can never give your child enough practice dividing fractions in the head. Here are some more. Feel free to add to these as you walk along. Remember, as you walk your dog, your cat, or your hamster, encourage your child to use what he or she knows to figure out these problems!

$8 \div \frac{1}{7}$ (56) $8\frac{2}{7} \div \frac{1}{7}$ (58) $8\frac{5}{7} \div \frac{1}{7}$ (61) $8\frac{6}{7} \div \frac{1}{14}$ (62)

$3 \div \frac{1}{6}$ (18) $3\frac{2}{6} \div \frac{1}{6}$ (20) $3\frac{1}{2} \div \frac{1}{6}$ (21) $3\frac{4}{6} \div \frac{1}{3}$ (11)

$2 \div \frac{1}{5}$ (10) $2\frac{1}{5} \div \frac{1}{5}$ (11) $2\frac{3}{5} \div \frac{1}{5}$ (13) $2\frac{4}{5} \div \frac{1}{10}$ (48)

On Your Own

FRACTIONS AT YOUR FINGERTIPS

Fractions are hard work, but it's well worth your time to learn to understand them. Life is filled with fractions, and because you have studied and practiced them, you are ready to participate in the world with one more tool in your pocket. See if you can find any real-life examples of division by fractions.

Now go have $\frac{1}{2}$ as good of a $\frac{16}{8}$ day!

Just for Fun!

Boy: Dad, can you help me find the lowest common denominator in this problem, please?

Dad: Don't tell me that they haven't found that yet! I remember looking for it when I was a boy.

Chapter 5

Decimals

In This Chapter

- Decimals: Friends to Fractions
- Decimal Draw
- Fractions to Decimals
- Decimals to Order
- Adding and Subtracting Decimals
- Multiplying and Dividing Decimals

DECIMALS: FRIENDS TO FRACTIONS

Parents' Corner

Like fractions, decimals name parts of a whole. Your child has already become a champion at fractions, so decimals should make sense. In decimal format, the whole is 1, and the decimal describes parts of that 1.

The root of *decimal*, *deci*, means "tenth," and in fact decimals come in tens. Decimals can describe tenths, hundredths, thousandths, or more. The further out the place value (the digit following the decimal point), the more precise the description. Decimals can get very small very quickly.

Here's what you probably already know about decimals: one-tenth = $\frac{1}{10}$ = .1. People say decimals several ways. For .1, some say "one-tenth"; others say "point one" (the point refers to the decimal point). .01 could be "one-hundredth" or "point zero one."

Decimals tend to make "cents" to kids for three reasons. First, decimals are used in money, and money usually motivates intermediate-age students. Second, for children who understand fractions and the relationship between fractions and decimals, decimals make a lot of sense. Finally, the decimal system is based on tens and their multiples—tens, hundreds, thousands, and so on—and children are often able to work with these multiples easily.

Teaching Tips

- ✓ Encourage your child to use what he or she knows about fractions and decimals to make the connections in this lesson's games.
- ✓ Use paper and pencil to illustrate examples.
- ✓ Dividing a numerator by a denominator to find a decimal is taught later.

The trick is to know which fractions and decimals say the same thing. Look over these to see the fraction/decimal connection:

$\frac{1}{10}$ = .1 $\frac{3}{10}$ = .3 $\frac{8}{10}$ = .8

$\frac{1}{100}$ = .01 $\frac{3}{100}$ = .03 $\frac{8}{100}$ = .08

$\frac{1}{1,000}$ = .001 $\frac{3}{1,000}$ = .003 $\frac{8}{1,000}$ = .008

AT THE KITCHEN TABLE

Conversion Concentration

To play this game, you need blank index cards and someone to play with. Choose which of the two versions of the game you would like to play (version 2 is more complex). Make the cards for that game. Note that some of the same decimals need to be on a few cards. This isn't a mistake: You will need these separate cards to play.

Version 1

$\frac{1}{10}$ $\frac{1}{3}$ $\frac{1}{4}$ $\frac{2}{10}$ $\frac{1}{5}$ $\frac{3}{10}$ $\frac{4}{10}$

$\frac{5}{10}$ $\frac{6}{10}$ $\frac{3}{5}$ $\frac{7}{10}$ $\frac{8}{10}$ $\frac{4}{5}$ $\frac{9}{10}$

.1 .333 .25 .2 .2 .3 .4

.5 .6 .6 .7 .8 .8 .9

Version 2

$\frac{2}{100}$ $\frac{1}{50}$ $\frac{4}{100}$ $\frac{1}{25}$ $\frac{5}{100}$ $\frac{1}{20}$ $\frac{8}{100}$ $\frac{2}{25}$

$\frac{4}{50}$ $\frac{2}{1,000}$ $\frac{1}{500}$ $\frac{4}{1,000}$ $\frac{2}{500}$ $\frac{1}{250}$ $\frac{5}{1,000}$ $\frac{1}{200}$

.02 .02 .04 .04 .05 .05 .08 .08

.08 .002 .002 .004 .004 .004 .005 .005

For either version, shuffle the cards and place them face down on the table in rows. When it is your turn, flip over two cards. Read the cards aloud. If you turned over two fractions or two decimals, your turn is over. Place the cards back where you found them, but remember where they are; you will need them to make future matches. If you turned over a decimal and a fraction, decide if they are equivalents. If they are equivalents, explain why to your opponent, keep the cards, and take another turn. If they are not equivalents, put them back where you found them and end your turn.

Keep playing until all the matches have been made. Count to see who has the most.

On Your Own

Talking Bread

Every basic fraction also has a decimal name. Let's talk bread, for example. Use paper and pencil to illustrate this problem as you work through it.

Imagine you just baked a loaf of bread. You and your nine hungry friends (10 all together) are waiting for a piece. To share the whole loaf evenly, you would cut the bread into 10 equal slices. Each friend would gladly take $\frac{1}{10}$, or .1, of the loaf. Your friends would be happy, and the bread would taste delicious.

Suppose you made raisin bread, and only you and one of your friends likes raisins. You already cut the bread into ten equal pieces, because you thought everyone wanted some. If you share the loaf evenly with your one friend, how much will you each get? It may help to draw a picture of a loaf of bread cut into 10 pieces. Then circle half of the pieces. You and your raisin-loving friend each get 5 of the 10 pieces. You each get $\frac{5}{10}$, or .5, of the bread. Look closely and you'll see that $\frac{5}{10} = \frac{1}{2}$. You each get half, or .5, of the whole loaf.

Now imagine there are four people who want to share the ten slices evenly—each person will get $\frac{1}{4}$ of the bread. How would you figure this out? Begin by sharing the bread slices as evenly as you can. Each person would get two-tenths of

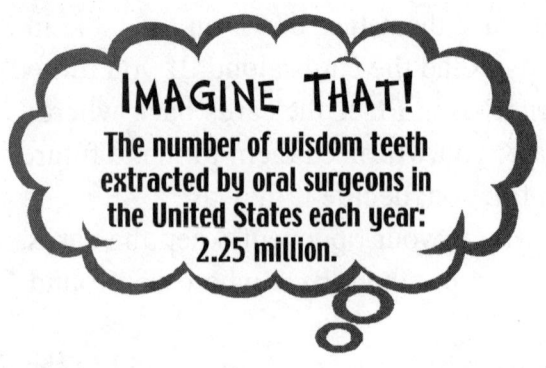

IMAGINE THAT!
The number of wisdom teeth extracted by oral surgeons in the United States each year: 2.25 million.

the bread, or .2, with two slices left over. Remembering that decimals are like tens, you cut the remaining slices into 10 pieces each. Draw this if it helps. Whenever you cut a tenth into 10 pieces, you have created hundredths. Pass out the hundredths to the four hungry and patient bread lovers. Each person will get 5 hundredths, or .05. Put all of this together, and each person has .2 + .05 = .25. Each of the four friends will get .25 of the whole loaf. $\frac{1}{4}$ = .25.

On Your Mark . . .

Watch a track event or a swimming competition or a downhill ski event on television. As the competitors race and finally cross the finish line or touch the wall, look at the competitor's time. Notice the decimals. For many sports, time is reported to the hundredth of a second. Can you imagine a hundredth of a second? That's pretty fast! Races that are extremely close require the breakdown of time to determine the winner. The athletes are also interested in their times. They always want to do better than they did before. Some athletes may even break times of previous athletes by hundredths of a second. Whew! This is tiring just to talk about!

On the Go...

CONVERSIONS

This game will help your child make the fraction/decimal connection quickly. While running errands, name a fraction or a decimal and have your child name the equivalent decimal (for a fraction) or fraction (for a decimal). If you say $\frac{1}{10}$, your child will say .1. If you say .5, your child will say $\frac{5}{10}$, or $\frac{1}{2}$. Here are some examples to ask your child.

$\frac{4}{10}$ (.4)   .25 ($\frac{25}{100}$, or $\frac{1}{4}$) $\frac{9}{10}$ (.9) $\frac{1}{3}$ (.333)   .6 ($\frac{6}{10}$, or $\frac{3}{5}$)   02 ($\frac{2}{100}$, or $\frac{1}{50}$)

$\frac{1}{2}$ (.5)   .333 ($\frac{1}{3}$)   .005 ($\frac{5}{1,000}$, or $\frac{1}{200}$)   .03 ($\frac{3}{100}$)   .008 ($\frac{8}{1,000}$, or $\frac{1}{125}$)

If your child stumbles over a decimal such as .002, have him or her count the place values (tenths, hundredths, thousandths). The place value furthest to the right is the denominator. The decimal number is the numerator: $.002 = \frac{2}{1,000}$, or $\frac{1}{500}$.

Challenge

Have your child use what he or she knows to figure out some more difficult conversions. If you asked, If $\frac{1}{4} = .25$, what would $\frac{3}{4}$ equal? Your child would need to figure out that $\frac{1}{4} + \frac{1}{4} + \frac{1}{4} = \frac{3}{4}$, so .25 + .25 + .25 = .75. Try these on your child:

If $\frac{1}{3} = .333$, what would $\frac{2}{3}$ equal? (.666)

If $\frac{4}{10} = .4$, what does $\frac{2}{5}$ equal? (Tell your child to use what he or she knows about equivalent fractions. Since $\frac{4}{10} = \frac{2}{5}$. The answer is $\frac{2}{5} = .4$.)

What is the decimal equivalent of $\frac{1}{20}$? ($\frac{1}{20} = \frac{5}{100}$ and $\frac{5}{100} = .05$, $\frac{1}{20} = .05$.)

Explain that anytime you have a fraction with a denominator that's not 10, 100, 1,000, or so on, see if it has an equivalent with a denominator of 10, 100, or 1,000. If so, rename it as a decimal. Try these:

$\frac{4}{5}$ ($\frac{8}{10} = .8$) $\frac{8}{20}$ ($\frac{4}{10} = .4$) $\frac{2}{200}$ ($\frac{1}{100} = .01$) $\frac{7}{70}$ ($\frac{1}{10} = .1$) $\frac{4}{40}$ ($\frac{1}{10} = .1$) $\frac{1}{5}$ ($\frac{2}{10} = .2$)

DECIMAL DRAW

Parents' Corner

Creating and seeing the size of a decimal in relation to a whole will help your child understand the size and value of the decimal. To get started, have your child draw a square and label it 1. Then have your child draw a decimal to draw—say .4—in another square. To do this, your child will have to know that he or she needs to divide the square into tenths and shade 4 of them. Now compare .4 and 1.

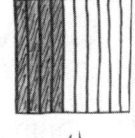

Think about a tile company that uses square tiles to tile floors. Suppose the company had to tile two square floors in an office building. In one day, the crew was able to tile one whole floor and .6 of the other. To explain the concept of decimals, draw an illustration to show that the crew tiled 1.6 of the floors and will do the other .4 of the floor tomorrow.

Another way to explain decimals is to have your child imagine a catering crew serving a birthday party. There are 100 guests, and you have made a square birthday cake. You would make 10 vertical cuts and 10 horizontal cuts to get 100 pieces. After putting piece after piece on plates, you are told to stop. Only 76 people wanted cake. The leftover cake in the pan is an example of decimals. You started with one entire cake. You now have 24 out of the original 100 pieces left over. In decimal-speak, .24 of the cake remains.

> **TEACHING TIPS**
>
> ✓ Have your child draw decimals to see their value and size and to practice figuring out decimal equivalents to fractions.
>
> ✓ Play the games several times—especially Practice, Practice, and More Practice. This game may be hard for some children to master.

AT THE KITCHEN TABLE

DECIMAL DRAWING

Since you have been converting some tough fractions to decimals, why not draw them to see how big they are? For this activity, you need scratch paper, a pencil, and a parent to play with.

Ask your parent to give you a decimal. Your job will be to draw the decimal, prove that you have drawn it correctly, and compare it to the number 1.

Here's an example: Your parent tells you .8. You smile and draw a square, divide it into tenths, and shade 8 of them. Tell your parent you know that you have .8 of the square shaded. Talk about how .8 is more than half of the square and is only .2 away from covering the whole square. Bravo! Here are some more decimals to draw:

1. .19 **2.** .33 **3.** .12 **4.** .89 **5.** .9 **6.** .44 **7.** .76 **8.** .99

There are no thousandths on this list. What would you have to do to draw the decimal .542? Aren't you *glad* there are no thousandths on this list?

On Your Own

PER 100

You know a lot about fractions and decimals, so percents will be easy for you. *Per cent* means "per 100," so any percent is a comparison to 100, just as decimals are comparisons to 100. The whole in percents is 100. If you get a 98 percent on your test, you have almost all of your answers right. If a car's gas tank is 15 percent full, it's getting pretty low—

Just for Fun!

What did one penny say to the other penny?

We make perfect cents!

imagine that the gas tank is made up of 100 little units, only 15 of them would be full.

Let's compare a fraction to a decimal to a percent.

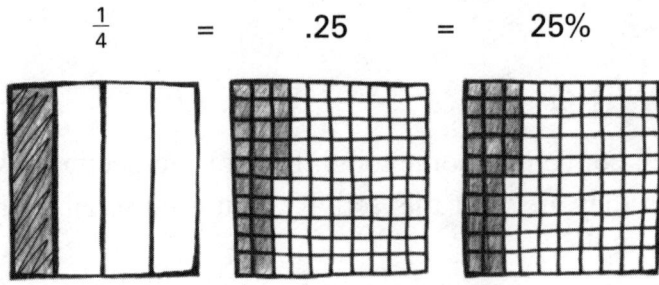

$\frac{1}{4}$ = .25 = 25%

Now try figuring out decimals and percents for the following fractions. Draw pictures on a separate sheet of paper to illustrate.

1. $\frac{1}{2}$ 2. $\frac{3}{4}$ 3. $\frac{1}{10}$

DECIMAL CONVERSION

Here's another chance to help your child practice switching some fractions to decimals. In this activity, give your child a fraction to be converted into a decimal. Note that the fractions might need to be converted to be renamed in decimal form.

Have your child name decimal equivalents of these fractions:

$\frac{8}{10}$ (.8) $\frac{2}{5}$ ($\frac{4}{10}$ = .4) $\frac{4}{25}$ ($\frac{16}{100}$ = .16) $\frac{3}{20}$ ($\frac{15}{100}$ = .15)

$\frac{4}{5}$ ($\frac{8}{10}$ = .8) $\frac{6}{50}$ ($\frac{12}{100}$ = .12) $\frac{12}{250}$ ($\frac{48}{1,000}$ = .048) $\frac{11}{500}$ ($\frac{22}{1,000}$ = .022)

FRACTIONS TO DECIMALS

Parents' Corner

Some decimal equivalents are easy to remember. Others will require some figuring out. Many fraction/decimal equivalents, in fact, have to be calculated. To find the more difficult decimal equivalents, simply do what the fraction tells you to do. $\frac{2}{5}$ tells you to divide 2 by 5. So set up the division equation and calculate:

```
       .4
    5)2.0
     - 20
        0
```

Therefore, $\frac{2}{5} = .4$

In some cases, the decimal repeats. Try $\frac{1}{7}$.

```
        .1428571
    7)1.0000000
     - 7
       30
     - 28
        20
      - 14
         60
       - 56
          40
        - 35
           50
         - 49
            10
          - 7
             30
           - 28
              2
```

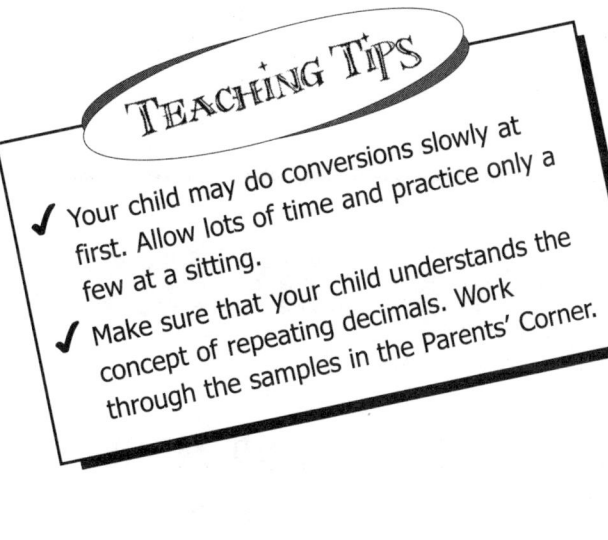

TEACHING TIPS

✓ Your child may do conversions slowly at first. Allow lots of time and practice only a few at a sitting.

✓ Make sure that your child understands the concept of repeating decimals. Work through the samples in the Parents' Corner.

Although the decimal took a long time to finally repeat, it will go on like this forever. A repeating decimal gets very close to an exact equivalent but never quite arrives. Sadly, a repeating decimal will never give you the exact fraction equivalent. To signify a repeating decimal, put a line over the entire portion that repeats. $\frac{1}{7} = .\overline{142857}$.

AT THE KITCHEN TABLE

DECIMAL DRILL

Practice is the surest way to become comfortable with converting fractions into decimals. Try one or two of these drills at a time. Take a break, run around the block, eat popcorn, or sing a song—and then try two more.

1. $\frac{5}{6}$
2. $\frac{8}{9}$
3. $\frac{7}{10}$
4. $\frac{2}{5}$
5. $\frac{3}{11}$
6. $\frac{5}{8}$
7. $\frac{1}{8}$
8. $\frac{3}{4}$

On the Go...

PRACTICE, PRACTICE, AND MORE PRACTICE

Here's a chance to help your child practice some fraction/decimal equivalents. Give your child a fraction to be converted into a decimal. In some cases, your child may need to find an equivalent fraction before renaming in decimal form. Here are some fractions to try:

$\frac{3}{5}$ ($\frac{6}{10}$ = .6) $\frac{8}{20}$ ($\frac{4}{10}$ = .4) $\frac{2}{25}$ ($\frac{8}{100}$ = .08) $\frac{1}{50}$ ($\frac{2}{100}$ = .02)

$\frac{9}{100}$ (.09) $\frac{42}{50}$ ($\frac{84}{100}$ = .84) $\frac{100}{250}$ ($\frac{4}{10}$ = .4) $\frac{90}{500}$ ($\frac{180}{1,000}$ = .180)

On Your Own

Cloudy and a Chance of Rain

Decimals are often used to describe various situations in more exact ways. Watch the weather report on your local news to hear a recent rainfall. Most likely, it will be given in hundredths of an inch. An umbrella will be your friend when there has been 1.50 inches of rain, but you might not notice .05 inch of rain. Most of us are used to seeing parts of inches in fractions. What fraction is the same as 1.50? What is the same as .05?

DECIMALS TO ORDER

Parents' Corner

Learning to change fractions to decimals is important, but your child's understanding of these concepts will be much stronger when he or she knows what the decimal stands for. Once your child knows the values of the decimals, he or she will have a much deeper sense of numbers. In this lesson, your child will practice ordering decimals by value, using what he or she knows about fractions. Drawings will illustrate the value of each decimal.

Decimals have value and meaning; they are the details of many stories. Use this example to explain the importance of decimals to your child.

Teaching Tips

- ✓ Let your child decide the pace of your practice. If practice seems easy, move quickly through the activities. If your child has difficulty doing and explaining activities, go slower.
- ✓ Use these activities as a way to talk about the value of decimals.
- ✓ If your child has trouble working through the race times in "Runners, Take Your Mark," review the different values of decimals. Sometimes comparing only two decimals at a time helps.

Imagine you are competing at a track meet, and it's your turn to race. As you crouch in the starting block, your heart is beating, and your mind is telling you that this will be your fastest race ever. You hear the call to get ready and then the blast: You're off. Your entire body works to get you to the finish line as fast as you can. But all of the racers seem to be crossing the finish line at once, so you eagerly watch for the times. When they come in, they're all decimals. Who was fastest, you wonder?

Once your child figures out how to arrange decimals, he or she will know who was fastest.

AT THE KITCHEN TABLE

DECIMAL PLOT

For this game, you need paper, pencil, a pile of index cards, and at least one other player. You and the other players will be plotting and ordering decimals. This will help you see how decimals compare to each other and to whole numbers.

Make two number cards with the whole numbers 33 and 34 on them. Place them far apart on a table top. These will be the two ends of your number line.

<-- 33 ————————————————————— 34 -->

Make a set of decimal cards. Write each of these decimals on a separate card.

33.50   33.17   33.12   33.543   33.1   33.9   33.99   33.611   33.61   33.8

Shuffle the cards. When it is your turn, take a decimal card and put it where you think it would go on the table between 33 and 34. If you pick 33.99, would that be closer to 33 or to 34? (Hint: Think of money!) If you need help, sketch the decimal portion of the number and label your drawing (you might need it later in the game to compare with another fraction). Explain your thinking to the other players, and show them your quick sketch. When they all give you the thumbs up, it's the next player's turn.

The next player turns over a decimal card and places it wherever it fits along the line with the 33, 33.99, and 34. Listen to each player's explanation so that you can agree or ask questions.

> **IMAGINE THAT!**
> You can tell the temperature in Fahrenheit by counting the number of clicks a cricket makes in 15 seconds and then adding 37.

Now, play the same game, but create new cards. You can make up cards of your own or use these whole numbers and decimals:

0	1	.5	.25	.75	.125
.33	.375	.66	.9	.1256	.91

HIGHER AND LOWER

Ask your parent to say a decimal. Name what two whole numbers the decimal is between. If your parent said 31.25, it is between 31 and 32. Then name a decimal that is higher than 31.25 that still falls between 31 and 32. Then name a decimal that is lower that still falls between 31 and 32. You might need paper and a pencil for this game so you can quick-sketch the decimals.

If your parent said 9.13, you would say it falls between 9 and 10. Now find a decimal that is lower than 9.13 that is still between 9 and 10. If you need to, quick-sketch the decimal part of the number.

.13 = 1-tenth and 3-hundredths.

What about 9.03? What would that look like? Quick-sketch only the decimal part (.03). Tell your parent why you think 9.03 is less than 9.13. Now you must find a decimal that is greater than 9.13 but still between 9 and 10. Maybe 9.99. Think this through and explain why 9.99 is greater than 9.13.

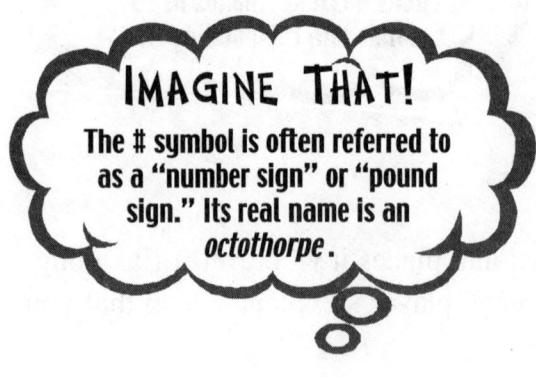

IMAGINE THAT!
The # symbol is often referred to as a "number sign" or "pound sign." Its real name is an *octothorpe*.

Higher and Lower Challenge

To make this game a little more difficult, your parent will say two simple words: "Get closer." In the previous sample, after you said 9.03, your parent would say, "Get closer." Your job now is to name a decimal that is closer to 9.13. Since 9.03 is close (only .1 away), you have to do some thinking. Use your quick sketches to help you name a closer decimal. 9.12! Your parent says, "Get closer." Hmm. This is not easy, but you have an idea. How about 9.125? Picture this decimal. It might be too tough to draw, but imagine a number line.

If you think of 9.12 and 9.13 on a number line, there are numbers that fall in between. Where would 9.125 fall? Right in the middle. Clear your throat, and say 9.125. Your parent will smile and say, "Get closer." Here you go again.

On Your Own

Runners, Take Your Mark

Imagine this: You've just had the race of your life, and your heart is still pounding. You look up to see that your time is 23.1 seconds. The other racers have come in with times of 25.25 seconds, 23.82 seconds, 24.01 seconds, and 23.11 seconds. What is the order of the racers, from slowest to fastest? Where did you place?

The first thing to do when ordering decimals is to look at the whole numbers. Put those in order first.

You will see that you have to do more to put the numbers that start with 23 in order. Of the three 23 numbers, which is the slowest speed? It may help to draw each decimal. (Don't worry about the whole number; everyone in this group ran for 23 seconds.) To compare 23.11 and 23.1, you may want to add a 0 to the end of 23.1. It's okay to add a 0 to the end of a decimal; it doesn't change the value. In fact, you could add an infinite number of zeros if you wanted and get a decimal that might look like this: 23.10000000000000. But why bother with all those zeros if you don't need them?

Who won the race?

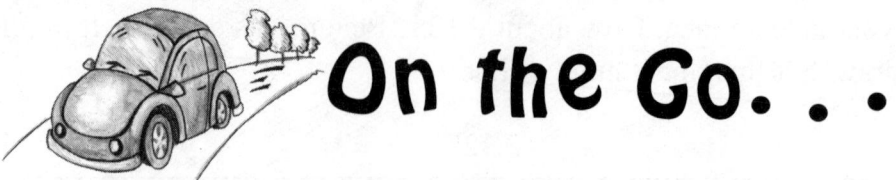

On the Go...

Traveling High and Low

After your child gets good with decimals, try playing Higher and Lower while on the go. You name a decimal. Your child tells you which two whole numbers the decimal falls between. You then ask your child to name another decimal that is still between those two whole numbers but that is either higher or lower than the original decimal. For example, if you said 2.5, your child would say that 2.5 falls between the whole numbers 2 and 3. If you ask your child to name a decimal higher than 2.5, he or she may come back with 2.9. Then ask for a number lower than 2.5. Your child may say 2.1. As a challenge, ask your child for a decimal number lower than 2.1. Once the concepts are understood, you can practice this while in the car, in a line, or in a waiting room somewhere.

ADDING AND SUBTRACTING DECIMALS

Parents' Corner

In this lesson, you are returning to a familiar place. Adding and subtracting decimals is easy for most children, even if they have never added or subtracted decimals before. They can use all of their prior knowledge to make this a successful mission, using the same steps as they use for whole-number addition and subtraction. The only thing to be careful about is lining up the decimals. Caution your child never to ignore that decimal point: It separates the whole numbers from the pieces. In decimal addition and subtraction, whole numbers are added to whole numbers and pieces are added to pieces. Just as in whole-number addition and subtraction, numbers may need to be carried over or borrowed, in this case from decimals to wholes.

Teach your child to estimate the answer first. A child who has not experienced drawing and ordering decimals might find it difficult to estimate that .93 + .98 will be close to 2. But with practice, your child will find that estimation is a great way to build the number sense required for mathematics. Every day, mathematicians rely on estimating and checking. Encourage your child to make estimation a habit—in all mathematics, but especially with fractions and decimals.

Teaching Tips

- ✓ Go through each kind of addition or subtraction problem with your child. Make up other examples if your child needs more practice.
- ✓ Emphasize estimation. Make a huge point of estimating answers before calculating them.
- ✓ Complete the activities with the goal of doing more and more decimal computing mentally.

At the Kitchen Table

Talking Problems

For this activity, you will need your parent, paper, a pencil, and some decimal problems to solve. Sit next to your parent. Copy down one of the problems from below. As you solve it, talk your way through every step so your parent can hear what you are thinking. Beware: Your parent may ask questions if any steps are unclear.

Here are the steps. Read through them and look at the addition problem.

```
  1 1
  62.98
+ 21.33
  —————
  84.31
```

> **Imagine That!**
> If you started at the number 1 and spelled out each number, you would get all the way to one thousand before the letter "a" appeared.

1. I estimate first. Add the whole numbers: 62 + 21 = 83. Scan the decimals: .98 is almost 1 whole, so the answer will be a little more than 84.

2. First I add the $\frac{8}{100}$ and the $\frac{3}{100}$. I get $\frac{11}{100}$, so I put down 1 and carry the other $\frac{10}{100}$, or $\frac{1}{10}$, to the tenth column.

3. Then I add the tenths column. $\frac{9}{10}$ and $\frac{3}{10}$ make $\frac{12}{10}$, plus the $\frac{1}{10}$ I carried. Now I have $\frac{13}{10}$. I put down the 3 and carry the other $\frac{10}{10}$, or 1 whole, to the 1s column.

4. I add the numbers in the ones column. 2 + 1 = 3, + 1 more from adding the tenths. I put 4 in the ones column.

5. Now I am ready to add the 10s column. 20 + 60 = 80, so I put 8 in the 10s column. My answer is 84.31. My estimate was a little more than 84, so I was right on track.

Get the idea? Just remember to estimate first and then talk your way through solving the problem. Here are a few to try. Make up more when you run out.

1. 77.21 − 52.78 = _____ **2.** 4.5 + 8.9 = _____

3. 16.56 + 8.65 = _____ **4.** 71.3 − 8.92 = _____

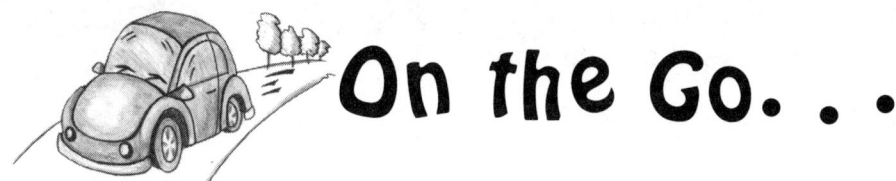

On the Go...

Decimals in Your Head

Your child can do anything—even add and subtract decimals without using pencil and paper. Give a few of these to try:

7 + 2.12 (9.12) 6 + 6.59 (12.59)

28 + 3.1 (31.1) 7 + 6 + 2.12 (15.12)

If your child has trouble solving these, remind him or her to add the whole numbers first and then tack on the decimals.

Now try subtraction.

10 − 2.30 (7.70)

For this one, remind your child to subtract the whole numbers first and then subtract the .30. Thinking of money might help. Try some more:

8 − 6.6 (1.4) 7 − 1.7 (5.3)

12 − 8.2 (3.8) 34 − 9.7 (24.3)

On Your Own

Decimal Estimation

This activity will help you practice estimating before adding and subtracting decimals. Start by working through a couple of problems, and then try more of your own. Before you begin, look at the example.

> Auntie Berry has $3.12 in her wallet. Her son repays her the $4.24 he borrowed to buy trading cards. How much money does she have now?

> $3.12
> + $4.24

Before you solve this, estimate about how much money you think Auntie now has. Look at the whole numbers and add those together in your head. Auntie has at least $7. Now scan the tenths and hundredths. Does she have enough to make another dollar? Does she have $0.50? No. Then she will have a little more than $7. Now solve the problem. 3.12 + 4.24 = $7.36. Your estimate was very close. Always estimate before calculating. This will help you develop your computing brain and figure out if you have the right or wrong answer.

Here's an example of subtraction.

> Auntie Berry needs to pay the milkman $2.25 out of the $7.36 she has in her wallet. How much does she now have in her wallet?

> $7.36
> − $2.25

Start by estimating. Subtract the whole dollar amounts in your head. $7 – $2 = $5. Now scan the cents. What is 40 – 30? Is this more or less than 50? Less, so $5 is a great estimate. Now calculate:

```
  $7.36
– $2.25
  $5.11
```
What a great estimator you are!

As in whole-number addition and subtraction, you may have to borrow or carry over in decimal equations. Try these:

```
  2.17         8.79
– 1.50       + 2.35
```

IMAGINE THAT!
The length of the beard an average man would grow if he never shaved: 27.5 feet. The number of inches whiskers grow per year: 5.5.

Estimate first. Look at the subtraction problem. The whole numbers are 2 and 1. What are you left with? Now scan the decimals to see if you have to regroup, or borrow and carry. Yes—so you will have less than 1. Compute. How was your estimate?

Look at the addition problem. Add the whole numbers first. Scan the decimals. Do you have enough to make one whole? Yes. Compute. How was your estimate?

You might also have equations where you will add or subtract a decimal from a whole number.

5 + 8.13

Add the whole numbers first. Now add the decimals. What do you get?

Try subtraction. Knowing how to subtract decimals from whole numbers will help when you are spending money.

You have $5.00 in your pocket. A package of pens costs $3.84, including tax. Do you have enough money? How much change will you get back?

Estimate first: $5 – $3 = $2. Now scan the cents. You still owe $.84, so your estimate goes down to a little more than a dollar.

Now compute.

$$\begin{array}{r} \$5.00 \\ -\$3.84 \\ \hline \end{array}$$

Borrowing across zeros can get old fast! Here's a fast strategy to use if you have a good memory. Turn $5.00 into $4.99. Don't forget about that penny. You'll need to add it back in when you are done.

$$\begin{array}{r} \$4.99 \\ -\$3.84 \\ \hline \$1.15 \\ +\$\ .01 \\ \hline \$1.16 \end{array}$$

$1.15   This is a snap to do—no regrouping!
+ $ .01   Don't forget to add the penny.
$1.16   Voila!

Try these:

1. $6.00 – $2.73 = _____   **2.** $10.00 – $6.55 = _____   **3.** $8.00 – $3.13 = _____

A Matter of Zeroes

With decimals, you can say the same thing in many ways. For example, 8.3 is the same as 8.30 and 8.300 and 8.3000000000000.

However, 8.3 is not the same thing as 8.300000001—although these numbers are very close. Which is bigger? By how much?

MULTIPLYING AND DIVIDING DECIMALS

Parents' Corner

Multiplying and dividing with decimals is related to whole-number computation. Your child must pay attention to the decimal. Practice with estimation will help your child predict a close range for the answer. For example, 14.2 × 3.7. The first step is to *estimate*. Quickly multiply the whole numbers in your head. (10 × 3) + (4 × 3) = 42. Your answer will be slightly bigger than 42. Now multiply, just as for a regular problem:

```
    14.2
  × 3.7       Ignore the decimal point for now.
    994
 + 4260
   5,254
```

Your multiplication is correct. Your estimate makes sense, but your answer does not. The explanation: When multiplying decimals, pay attention to the decimals. Count the numbers after the decimal point in the original problem. There are two numbers after the decimal points (one for each factor). Count back two places from the end of the number in your answer and put the decimal point there. Your answer is 52.54, which is close enough to your estimate to be reasonable.

> **TEACHING TIPS**
> - ✓ Have your child estimate all answers first.
> - ✓ Compare the estimate with the answer. If the digits are the same, but the answer seems to high or low, look at the decimal portion of the problem. If the digits are very different, check your calculations.
> - ✓ When first doing division, set up the problems for your child.

Now look at 104 ÷ 3.4. First estimate. 104 ÷ 3 is about 34. Your answer will be around 34—not 340 or 3,400. Move the decimal point to the end of the divisor. *Remember:* Whatever you do to one side must be done to the other so you do not change the problem. If you move the decimal one place to the right in the divisor, you must also move it one place to the right in the dividend by adding a zero:

34)1,040.

The rest is easy division!

AT THE KITCHEN TABLE

TALKING PROBLEMS (AGAIN)

For this activity, you will need your parent, paper, a pencil, and some decimal problems to solve. Sit side by side with your parent. Copy one of the problems from below. Estimate your answer first and then write down your estimation. As you solve the problem, talk your way through every step so your parent can hear your thinking. Compare your estimate to your final answer. Beware: Your parent may ask questions if any of your steps seems unclear.

1. 99.3×8.3
2. 22.44×3.2
3. $12.3 \div 2.7$
4. $48 \div 6.1$

On Your Own

BALLPARK ESTIMATE

Whenever you multiply decimals, ignore the decimal point until you are at your final answer. Before computing, estimate what you think the answer will be. For 5.4×8.3, estimate by multiplying the whole numbers first. $5 \times 8 = 40$. Now multiply the decimals. $4 \times 3 = 12$, so the final answer will be right around 41. The real answer is 44.82. Not bad! Try a few of these:

1. 9.6×7.8
2. 200×2.16
3. $33.7 \times .961$

Don't forget to estimate first! Then compute. Are your answers close to your estimations? If not, figure out why not and try again.

Now try some division problems. Look at this one:

7.2)28

Your first job is to estimate. Look at the whole numbers: 28 ÷ 7 = 4. Your answer will be around 4—not 40 or 400, just 4.

Your second job is to get the decimal point to the end of the divisor. Mathematics is about balance. Whatever you do to one side, you need to do to the other so you do not change the problem. Since you move the decimal one place to the right in the divisor, you must also move it one place in the dividend. There is no place to go in the dividend, so add a zero. When you have moved the decimal point out of the divisor, put another decimal point directly above where it is in the dividend for your answer.

72)280.

Now estimate these problems:

4. 3.6)80.23

5. 14)99.8

6. 2.25)90.6

Set up your problems for division. Do they look like this?

36)802.3   14)99.8   225)9060.

Bravo! Now divide. If long division is hard for you, check the division section of this book to find another way of dividing that might help.

Compare your answer to your estimate. Are you in the ballpark?

Money, Money, Money

Everyone likes money. And how is money expressed? You're right—in decimals! If you have 10 cents in your pocket, you have $\frac{1}{10}$ of a dollar, or .10.

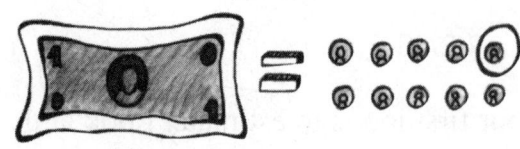

If you have $3.50, you have 3.50 dollars, or $3\frac{1}{2}$ dollars. How much would .25 of a dollar be? $\frac{25}{100}$. What coin represents this amount? What *coins* could represent this amount?

.25 = 1 quarter

.25 = 2 dimes and a nickel, or .10 + .10 + .05

Can you think of other ways to get .25 of a dollar?

Go back to the dollar. If a dollar is a whole, or 1, then $\frac{1}{100}$ of a dollar is .01, or 1 cent.

1. How much money is .05, or 5 hundredths, of a dollar? _____

2. How about .25, or 25 hundredths, of a dollar? _____

3. .19 of a dollar? _____

Now suppose 4 children in a family all contribute .25 of a dollar to purchase a Mother's Day present. How much money would they have? What if each child contributed $0.50? What if the 4 siblings and their 3 cousins each contributed $0.50 to purchase a present for Grandma?

Now suppose Grandma has $14.00. She wants to split the money evenly among her 7 grandchildren. How much does each grandchild get?

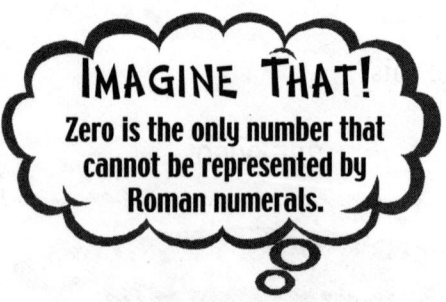

IMAGINE THAT! Zero is the only number that cannot be represented by Roman numerals.

On the Go...

ESTIMATE, ESTIMATE

Because estimating will help your child determine whether he or she has arrived at a reasonable answer, it's a good idea to practice it. Next time you are raking leaves or pulling weeds, call out some decimal multiplication and division problems. Your child doesn't have to solve them, but should get a good estimate and explain the estimate. For example, you might say, 7.3 × 2.9. Your child may estimate that 7 × 2 = 14, so the answer will be greater than but close to 14. Or your child might round up and say, 7 × 3 = 21. This is a closer estimate. Call out other problems, such as

4.88 × 100 (about 500)

3.976 × 8 (about 32)

17.1 × 1.15 (a little more than 17)

Estimating in division is good to practice, too. For example, 20.5 ÷ 5.2. Your child may estimate 20 ÷ 5 = 4. So the answer will be around 4. Try some problems like these:

8.3 ÷ 2.22 (about 4)

111 ÷ 10.1 (about 11)

63.6 ÷ 7.1 (about 9)

Answers

CHAPTER 1

Strategies Strategies Strategies, pp. 6–8
× 0: 7 × 0 = 0; 43 × 0 = 0; 3,924 × 0 = 0
× 1: 7 × 1 = 7; 43 × 1 = 43; 3,924 × 1 = 3,924
× 2: 4 × 2 = 8; 6 × 2 = 12; 8 × 2 = 16
× 3: 4 × 3 = 12; 5 × 3 = 15; 8 × 3 = 24
× 4: 4 × 4 = 16; 5 × 4 = 20; 8 × 4 = 32
× 5: 4 × 5 = 20; 6 × 5 = 30; 7 × 5 = 35
× 6: 3 × 6 = 18; 4 × 6 = 24; 8 × 6 = 48
× 8: 4 × 8 = 32; 6 × 8 = 48; 9 × 8 = 72
× 9: 4 × 9 = 36; 8 × 9 = 72; 9 × 9 = 81

The Trick of Tens, p. 14
1. 600; **2.** 2,300; **3.** 8,200; **4.** 8,000; **5.** 9,000,000; **6.** 120,000

Fancy Multiplications, p. 16
1. 16; **2.** 160; **3.** 1,600; **4.** 1,600; **5.** 160,000; **6.** 27; **7.** 270; **8.** 2,700; **9.** 2,700; **10.** 270,000; **11.** 28; **12.** 280; **13.** 2,800; **14.** 2,800; **15.** 280,000

CHAPTER 2

Gazinta, pp. 34–36
÷ 2: All are divisible by 2.
÷ 3: 27, 42, 105 are divisible by 3.
÷ 4: 336, 572, 188 are divisible by 4.
÷ 5: 115, 360, 85 are divisible by 5.
÷ 6: 24, 48, 66 are divisible by 6.
÷ 8: 6,832, 3,264 are divisible by 8.
÷ 9: 243, 468, 522 are divisible by 9.
÷ 10: 440, 600 are divisible by 10.

It's All in the Head, p. 41
1. 40; **2.** 4; **3.** 40; **4.** 500; **5.** 50; **6.** 5; **7.** 800; **8.** 80; **9.** 8; **10.** 7; **11.** 70; **12.** 700

Answers

Daddy, Mommy, Sister, Brother, p. 43

1.
```
       66 R 9
   13)867
     - 78
       87
     - 78
        9
```

2.
```
       14 R 11
   37)529
     - 37
      159
    - 148
       11
```

3.
```
         57 R 8
   24)1,376
     - 120
       176
     - 168
         8
```

4.
```
         795 R 2
   12)9,542
     - 84
      114
    - 108
       62
     - 60
        2
```

Multiplication in Reverse, p. 45

1.
```
       66   R 9
   13)867
     - 130   10
       737
     - 130   10
       607
     - 130   10
       477
     - 130   10
       347
     - 130   10
       217
     - 130   10
        87
     -  65    5
        22
     -  13    1
         9
```

2.
```
       14   R 11
   37)529
    - 370   10
      159
    - 148    4
       11
```

3.
```
         57   R 8
   24)1,376
    - 1,200   50
       176
     -  120    5
         56
     -   48    2
          8
```

4.
```
         795   R 2
   12)9,542
    - 8,400   700
      1,142
    - 1,080    90
         62
     -   60     5
          2
```

128 Answers

CHAPTER 3

Sandwiches to Go, p. 50
Each person will get $\frac{3}{4}$ of a sandwich.

Group Parts, p. 53
8 bicycles = 4; 3 cookies = $1\frac{1}{2}$; 10 gifts = 5; 9 apples = $4\frac{1}{2}$; 4 coats = 2; 1 banana = $\frac{1}{2}$; 2 lollipops = 1; 5 sandwiches = $2\frac{1}{2}$; 6 basketballs = 3; 7 strawberries = $3\frac{1}{2}$; 11 crackers = $5\frac{1}{2}$; 12 crayons = 6

Crack the Code, p. 58
1. 16; **2.** 4; **3.** 18; **4.** 2; **5.** 3; **6.** 6

Mixed Sandwiches, p. 62
$\frac{7}{4}$ = 1 whole sandwich and 3 fourths of a sandwich

The Half Shuffle, p. 65
5—numerator: $\frac{5}{10}$; 6—denominator: $\frac{3}{6}$; 6—numerator: $\frac{6}{12}$
7—numerator: $\frac{7}{14}$; 8—numerator: $\frac{8}{16}$; 8—denominator: $\frac{4}{8}$
9—numerator: $\frac{9}{18}$; 10—numerator: $\frac{10}{20}$; 10—denominator: $\frac{5}{10}$
2—numerator: $\frac{2}{4}$; 2—denominator: $\frac{1}{2}$; 3—numerator: $\frac{3}{6}$
4—numerator: $\frac{4}{8}$; 4—denominator: $\frac{2}{4}$; 12—denominator: $\frac{6}{12}$

The Quarter Shuffle, p. 66
2—numerator: $\frac{2}{8}$; 3—numerator: $\frac{3}{12}$; 4—numerator: $\frac{4}{16}$
4—denominator: $\frac{1}{4}$; 5—numerator: $\frac{5}{20}$; 6—numerator: $\frac{6}{24}$
7—numerator: $\frac{7}{28}$; 8—numerator: $\frac{8}{32}$; 8—denominator: $\frac{2}{8}$
9—numerator: $\frac{9}{36}$; 10—numerator: $\frac{10}{40}$; 11—numerator: $\frac{11}{44}$
12—denominator: $\frac{3}{12}$; 16—denominator: $\frac{4}{16}$; 20—denominator: $\frac{5}{20}$

The Third Shuffle, p. 66
2—numerator: $\frac{2}{6}$; 3—numerator: $\frac{3}{9}$; 3—denominator: $\frac{1}{3}$
4—numerator: $\frac{4}{12}$; 5—numerator: $\frac{5}{15}$; 6—numerator: $\frac{6}{18}$
6—denominator: $\frac{2}{6}$; 7—numerator: $\frac{7}{21}$; 8—numerator: $\frac{8}{24}$
9—numerator: $\frac{9}{27}$; 9—denominator: $\frac{3}{9}$; 10—numerator: $\frac{10}{30}$
12—denominator: $\frac{4}{12}$; 15—denominator: $\frac{5}{15}$; 18—denominator: $\frac{6}{18}$

Answers

Fraction Draw, p. 68

1.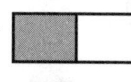
2.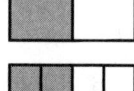
3.

Two (or More) of the Same, p. 74

1. Answers may include $\frac{2}{4}, \frac{3}{6}, \frac{4}{8}, \frac{5}{10}, \frac{6}{12}, \frac{7}{14}, \frac{8}{16}, \frac{9}{18}, \frac{10}{20}, \frac{11}{22}$ and any other equivalent of $\frac{1}{2}$.
2. Answers may include $\frac{2}{8}, \frac{3}{12}, \frac{4}{16}, \frac{5}{20}, \frac{6}{24}, \frac{7}{28}, \frac{8}{32}, \frac{9}{36}, \frac{10}{40}, \frac{11}{44}$ and any other equivalent of $\frac{1}{4}$.
3. Answers may include $\frac{2}{6}, \frac{3}{9}, \frac{4}{12}, \frac{5}{15}, \frac{6}{18}, \frac{7}{21}, \frac{8}{24}, \frac{9}{27}, \frac{10}{30}, \frac{11}{33}$ and any other equivalent of $\frac{1}{3}$.
4. Answers may include $\frac{6}{8}, \frac{9}{12}, \frac{12}{16}, \frac{15}{20}, \frac{18}{24}, \frac{21}{28}, \frac{24}{32}, \frac{27}{36}, \frac{30}{40}, \frac{33}{44}$ and any other equivalent of $\frac{3}{4}$.

CHAPTER 4

Building Problems, p. 78

1. $\frac{3}{4} + \frac{3}{4} = \frac{6}{4} - \frac{1}{4} = \frac{5}{4}$, or $1\frac{1}{4}$
2. $\frac{3}{8} + \frac{5}{8} = \frac{8}{8} + \frac{7}{8} = \frac{15}{8} - \frac{1}{8} = \frac{14}{8}$, or $1\frac{6}{8}$, or $1\frac{3}{4}$
3. $\frac{2}{3} + \frac{2}{3} = \frac{4}{3} + \frac{4}{3} = \frac{8}{3}$, or $2\frac{2}{3}$
4. $8 + \frac{1}{4} = 8\frac{1}{4} + 1\frac{1}{4} = 9\frac{2}{4}$, or $9\frac{1}{2}$
5. $\frac{1}{4} + \frac{1}{4} + \frac{1}{4} = \frac{3}{4} + \frac{3}{4} = \frac{6}{4} + 1\frac{3}{4} = 1\frac{9}{4}$, or $3\frac{1}{4}$
6. $\frac{5}{8} + \frac{5}{8} = \frac{10}{8} + 2 = 2\frac{10}{8} - 1\frac{1}{8} = 1\frac{9}{8}$, or $2\frac{1}{8}$

Cooking with Fractions, pp. 81–82

$\frac{3}{4} \times 2 = \frac{6}{4}$, or $1\frac{1}{2}$ cups butter

$1\frac{1}{2} \times 2 = \frac{3}{2} \times 2 = \frac{6}{2}$, or 3 cups brown sugar

$1\frac{3}{4} \times 2 = \frac{7}{4} \times 2 = \frac{14}{4}$, or $3\frac{1}{2}$ cups flour

$\frac{2}{3} + \frac{3}{4} = \frac{8}{12} + \frac{9}{12} = \frac{17}{12}$, or $1\frac{5}{12}$ cup butter

$1\frac{1}{2} + 1\frac{3}{4} = 1\frac{2}{4} + 1\frac{3}{4} = 2\frac{5}{4}$, or $3\frac{1}{4}$ cups brown sugar

Half a Fraction, p. 85

Check drawings to be sure they make sense.

1. 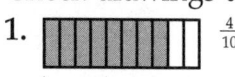 $\frac{4}{10}$
2. 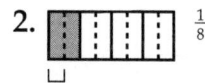 $\frac{1}{8}$
3. 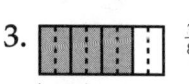 $\frac{3}{8}$
4. 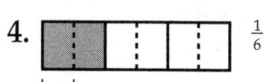 $\frac{1}{6}$

5. 1/12   6. 1/20

7. 1/5   8. 1/8

9. 1/4   10. 1/9

11. 1/6   12. 1/24

Shrinking Numbers, p. 86
1. $\frac{1}{8}$; 2. $\frac{1}{12}$; 3. $\frac{1}{4}$

Simply Fractions, p. 89
1. $\frac{1}{2}$; 2. $\frac{3}{4}$; 3. $\frac{3}{4}$; 4. $\frac{2}{3}$

Math Wiz, p. 93
1. $2 \div \frac{1}{4} = 8$; $2\frac{1}{2} \div \frac{1}{4} = 10$; $2\frac{4}{8} \div \frac{1}{4} = 10$
2. $\frac{4}{8} \div \frac{1}{4} = 2$; $\frac{5}{8} \div \frac{1}{4} = 2\frac{1}{2}$; $\frac{3}{8} \div \frac{1}{4} = 1\frac{1}{2}$
3. $\frac{6}{8} \div \frac{1}{8} = 6$; $\frac{4}{8} \div \frac{1}{8} = 4$; $\frac{1}{2} \div \frac{1}{8} = 4$
4. $\frac{1}{4} \div \frac{1}{8} = 2$; $\frac{3}{4} \div \frac{1}{8} = 6$; $1\frac{1}{2} \div \frac{1}{8} = 12$

Flip and Multiply, pp. 95–96
Check that estimates make sense and that work is shown.
1. $\frac{3}{4} \div \frac{1}{3} = \frac{3}{4} \times \frac{3}{1} = \frac{9}{4}$, or $2\frac{1}{4}$
2. $1\frac{2}{3} \div \frac{4}{5} = \frac{5}{3} \div \frac{4}{5} = \frac{5}{3} \times \frac{5}{4} = \frac{25}{12}$, or $2\frac{1}{12}$
3. $\frac{7}{8} \div \frac{1}{3} = \frac{7}{8} \times \frac{3}{1} = \frac{21}{8}$, or $2\frac{5}{8}$
4. $\frac{8}{9} \div \frac{3}{4} = \frac{8}{9} \times \frac{4}{3} = \frac{32}{27}$, or $1\frac{5}{27}$
5. $1\frac{4}{5} \div \frac{1}{8} = \frac{9}{5} \div \frac{1}{8} = \frac{9}{5} \times \frac{8}{1} = \frac{72}{5}$, or $14\frac{2}{5}$
6. $\frac{6}{11} \div \frac{2}{5} = \frac{6}{11} \times \frac{5}{2} = \frac{30}{22}$, or $1\frac{8}{22}$, or $1\frac{4}{11}$
7. $2\frac{2}{3} \div \frac{3}{8} = \frac{8}{3} \div \frac{3}{8} = \frac{8}{3} \times \frac{8}{3} = \frac{64}{9}$, or $7\frac{1}{9}$
8. $\frac{9}{10} \div \frac{2}{3} = \frac{9}{10} \times \frac{3}{2} = \frac{27}{20}$, or $1\frac{7}{20}$
9. $\frac{6}{8} \div \frac{1}{7} = \frac{6}{8} \times \frac{7}{1} = \frac{42}{8}$, or $5\frac{2}{8}$, or $5\frac{1}{4}$

CHAPTER 5

Decimal Drawing, p. 105

1.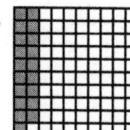
2.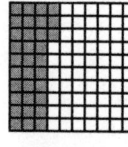
3.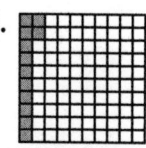
4.
5.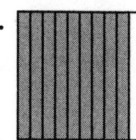
6.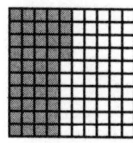
7.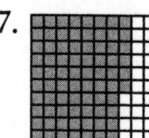
8.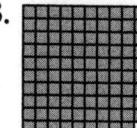

Per 100, p. 106

1. $\frac{1}{2} = .5 = 50\%$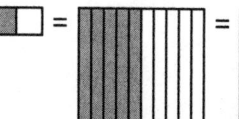

2. $\frac{3}{4} = .75 = 75\%$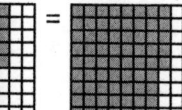

3. $\frac{1}{10} = .1 = 10\%$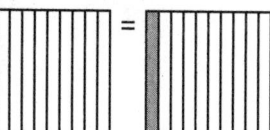

Decimal Drill, p. 108

1.
```
    .833
6)5.00
  −48
   20
  −18
    2
```

2.
```
    .888
9)8.000
  −72
   80
  −72
   80
  −72
    8
```

3.
```
    .7
10)7.0
  −70
    0
```

4.
```
   .4
5)2.0
 −20
   0
```

5.
```
        .2727
    11)3.0000
       - 22
         80
       - 77
         30
       - 22
         80
       - 77
          3
```

6.
```
       .625
    8)5.000
     - 48
        20
      - 16
        40
      - 40
         0
```

7.
```
       .125
    8)1.000
     - 8
       20
     - 16
       40
     - 40
        0
```

8.
```
       .75
    4)3.00
     - 28
       20
     - 20
        0
```

Cloudy and a Chance of Rain, p. 109
$.50 = 1\frac{1}{2}$, or $\frac{3}{2}$
$.05 = \frac{5}{100}$, or $\frac{1}{20}$

Decimal Plot, pp. 111–112
33, 33.1, 33.12, 33.17, 33.50, 33.543, 33.61, 33.611, 33.8, 33.9, 33.99, 34
0, .125, .1256, .25, .33, .375, .5, .66, .75, .9, .91, 1

Runners Take Your Mark, p. 113
25.25, 24.01, 23.82, 23.11, 23.1; You placed first!

Talking Problems, p. 117
1.
```
       6 11 1
      77.2̸1̸
    - 52.78
      24.43
```

2.
```
        1
       4.5
     + 8.9
      13.4
```

3.
```
      1 1 1
     16.56
    + 8.65
     25.21
```

4.
```
      10 12
     6 0̸ 2̸0̸
    71.3̸0̸
   - 8.92
    62.38
```

Decimal Estimation, p. 120
1. $6.00 − $2.73 = 5.99 − 2.73 = 3.26 + .01 = $3.27
2. $10.00 − $6.55 = 9.99 − 6.55 = 3.44 + .01 = $3.45
3. $8.00 − $3.13 = 7.99 − 3.13 = 4.86 + .01 = $4.87

A Matter of Zeroes, p. 120
8.300000001 is bigger, by .000000001.

Talking Problems (Again), p. 122

1. Estimate: 100 × 8 = about 800
Answer: 824.19
2. Estimate: 22 × 3 = about 66
Answer: 71.808
3. Estimate: 12 ÷ 3 = about 4
Answer: 4.$\overline{555}$
4. Estimate: 48 ÷ 6 = about 8
Answer: 7.869

Ballpark Estimate, pp. 122–123

1. Estimate: 10 × 8 = about 80

```
    4
   9.6
 × 7.8
 ─────
   768
  1
 + 672
 ─────
 74.88
```

2. Estimate: 200 × 2 = about 400

```
   200
 × 2.16
 ─────
  1200
   200
 + 400
 ──────
 432.00
```

3. Estimate: 30 × 1 = about 30

```
    33.7
  × .961
  ──────
     337
    2022
  + 3033
  ──────
  32.3857
```

4. Estimate: 80 ÷ 4 = about 20

```
        22.28611
    36)802.30000
       − 72
       ────
         82
        − 72
        ────
         103
        − 72
        ────
         310
       − 288
       ─────
          220
        − 216
        ─────
           40
         − 36
         ────
           40
```

5. Estimate: 100 ÷ 10 = about 10

```
        7.12857 . . .
   14)99.80000
      − 98
        ──
        18
      − 14
        ──
         40
       − 28
         ──
         120
       − 112
         ───
           80
         − 70
           ──
           100
          − 98
            ──
            20
```

6. Estimate: 90 ÷ 2 = about 45

```
         40.266
   225)9060.0
      − 900
        ───
         60
       − 00
         ──
         600
       − 450
         ───
         1500
       − 1350
         ────
         1500
```

Money, Money, Money, p. 124

1. 5 cents, or 1 nickel, or 5 pennies
2. 25 cent, or 1 quarter, or 2 dimes and 1 nickel, or 5 nickels, or 25 pennies, etc.
3. 19 cents, or 1 dime, 1 nickel, and 4 pennies, etc.

If all 4 children contribute .25, they would have $1.00 total.
If all 4 children contribute .50, they would have $2.00 total.
If 4 children and their 3 cousins contribute $.50, they would have $3.50.
If Grandma split $14.00 among 7 grandchildren, each grandchild would get $2.00.

Index

"Adventure Math," 22
Algorithm, 42, 43, 72, 87, 94
"Alligator Munch," 70
"Are You First?", 71
Array, 25, 26–27

"Ballpark Estimate," 122–123
Building problems, 77–78, 80; common denominator, 77–78; mixed numbers with like denominators, 78

"Cloudy and a Chance of Rain," 109
Commutative property of multiplication, 2
Composite numbers, 24, 26, 27
Compute numbers, 5
Computing, 39
"Conversion Concentration," 100–101
"Conversions," 103
"Cooking with Fractions," 81–82
"Crack the Code," 58
Crunching numbers, 5

"Daddy, Mommy, Sister, Brother," 42, 43. See also Long division
"Decimal Conversion," 106
"Decimal Drawing," 105
"Decimal Drill," 108
"Decimal Estimation," 118–120. See also Estimation
"Decimal Plot," 111–112
Decimals, 99, 104, 107, 110, 115, 121; adding, 115–120; dividing, 121–123, 125; drawing, 104–105, 110; exercises/problems, 106, 108, 117, 124, 125; games/activities, 100–101, 103, 105, 111–112, 114, 116–120, 122–123; money and, 124; multiplying, 121, 122, 125; ordering, 110–114; subtracting, 115–120; teaching tips, 99, 104, 107, 110, 115, 121; weather report and, 109; zeroes and, 120. See also Fraction/decimal equivalents
"Decimals in Your Head," 117
Denominator, 28, 48, 55, 56, 60, 64, 65, 66, 67, 69, 72, 87, 89. See also Mixed denominators
"Dice Multiplice," 3
Dividend, 33, 42, 90, 94, 123
"Divisibility Number Cruncher," 36–37
Division, 33, 39, 42; algorithms, 42, 43; exercises/problems, 37, 41, 43, 45–46; games/activities, 36–37, 38, 40–41; "gazinta" rules, 24, 34–36; multiplication and, 33, 39; practicing, 38; subtraction and, 33; teaching tips, 33, 39, 42; vocabulary, 33. See also Fractions, dividing; Long division
"Division Concentration," 40
Division facts, 39
Divisor, 33, 42, 94, 123

Equivalent, definition of, 72
"Estimate and Divide," 46
Estimation, 20–23, 39, 46, 94, 115, 118–120, 121, 122–123; making routine, 20; practicing, 38, 125; skill drills, 23. See also Front-end estimating; Rounding

Factor, 2, 24. See also Greatest common factor
"Factor Fiction," 30
"Factor Figures," 25–26
"Factor Find," 27
"Fancy Multiplications," 16
"Flip and Multiply," 95–96
Foreign math, 46
"Fraction Challenge," 51
"Fraction Concentration," 57–58
Fraction/decimal equivalents, 107
"Fraction Draw," 68
"Fraction Illustration," 49–50
Fractions, 48, 52, 55, 60, 64, 69, 72, 76, 83, 87, 90, 94; basic, 64, 67, 76; counters and, 55; drawings and, 55, 69; equivalent, 64, 72, 74; exercises/problems, 51, 54, 58–59, 68, 71, 77–78, 80, 86, 89, 92–93, 95–97; games/activities, 49–50, 53–54, 56–58, 61–63, 65–67, 70–71, 73–74, 78–80, 81–82, 84–85, 88, 91–92; ordering, 69, 71; part-to-whole relationships, 52; simplest form, 89; teaching tips, 48, 52, 55, 60, 64, 69, 72, 76, 83, 87, 90, 94. See also Building problems; Denominator; Fractions, adding; Fractions, dividing; Fractions, multiplying; Fractions, subtracting; Improper fractions; Mixed numbers, adding; Mixed numbers, subtracting; Numerator
Fractions, adding, 53, 76–82; practice in kitchen, 81–82; with common denominators, 76; with unlike denominators, 76
Fractions, dividing, 90–97; drawing problems, 90; real-life examples, 97. See also Invert and multiply rule
Fractions, multiplying, 83–89; drawing problems, 83; multiplication sign meaning, 83; rule for, 87
Fractions, subtracting, 76–82; practice in kitchen, 81–82; with common denominators, 76; with unlike denominators, 76
"Fractions in Disguise," 56
Fraction-speak, 48
"Fraction Workout," 74
Front-end estimating, 20

"Gazinta, Divide, and Draw," 91–92
"Gazinta," 34–36
"GCF Concentration," 29

Index

Greatest common factor (GCF), 24, 28, 30
"Group Parts," 53

"Half a Fraction," 84–85
"Half Shuffle," 65
"Half Truths," 54
Head problems, 80, 93, 97, 117. *See also* Mental computation
"Higher and Lower," 112–113, 114
"How Many," 63

Identity Property of Multiplication, 6, 10
Improper fractions, 60, 76, 88; multiplying mixed numbers and, 87
Inverse, 33, 39
Invert-and-multiply algorithm, 90, 94, 95; practicing, 95–96
"It's All in the Head," 41

Long division, 42–46; "Divide, Multiply, Subtract, Bring down" method, 42, 43, 44; exercises/problems, 44–45; "Multiplication in Reverse" method, 42

Manipulatives, 90
Math sense, 94
"Math Wiz," 93
"Matter of Zeroes," 120
Mental computation, 17, 28, 87, 89, 115; as mathematically efficient, 5. *See also* Head problems; "Why Bother?"
"Mental Multiplication," 19
Mixed denominators, ordering fractions with, 71
Mixed fractions, 88

Mixed numbers, 60, 76; building problems, 78; dividing, 92. *See also* Mixed numbers, adding; Mixed numbers, multiplying; Mixed numbers, subtracting
Mixed numbers, adding, 76; with common denominators, 76; with unlike denominators, 76
Mixed numbers, multiplying, 87; improper fractions and, 87
Mixed numbers, subtracting, 76; with common denominators, 76; with unlike denominators, 76
"Mixed Sandwiches," 62, 128
"Money, Money, Money," 124
Multiplicand, 21
Multiplication, 2, 5, 13, 17, 20, 24, 28; addition and, 2; division and, 33, 39; flexibility of, 2; games/activities, 3–4, 9, 12, 14–15, 18, 19, 21–22, 25–27, 29–31; exercises/problems, 6–8, 10–11, 14, 16, 18–19, 22–23; simplifying equations, 17; teaching tips, 2, 5, 13, 17, 20, 24, 28; terms, 2; two-digit, 17. *See also* "Fancy Multiplications;" Fractions, multiplying
"Multiplication Concentration," 14–15
Multiplication fact chart, 10–11
Multiplication facts, 4, 5; basic, 13, 17; basic-facts-in-hiding, 13; effort required for learning, 5; memorization of, 5; time required for learning, 5; understanding, 5
"Multiplication Mystery," 18
Multiplication strategies, 5, 6–8

"Name That Half," 67
"Number Cruncher," 9
Number crunching machine, 9, 36. *See also* T-chart
Number sense, vii; estimation and, 20. *See also* Math sense
Numerator, 48, 55, 56, 60, 64, 65, 66, 67, 69, 72, 87, 89

Percent, 105–106
"Per 100," 105–106
"Pigs and Chickens," 21–22
Powers of 10: dividing by, 39; multiplying by, 13, 16, 17
"Practice, and Practice More," 88
"Practice, Practice, and More Practice," 104, 108
Prime numbers, 24, 26
Product, 2, 24, 42

"Quarter Shuffle," 66
"Quick Draw," 61
Quotient, 33, 90

Remainder, 44, 45
Repeating decimals, 107
Repetition, memorizing multiplication facts with, 12
Rounding, 20
"Runners, Take Your Mark," 110, 113

"Sandwiches to Go," 50
"Shrinking Numbers," 86
"Skinny and Plump," 26–27

Skip-counting, 3–4
"Skip-Counting Calisthenics," 4
"Spin and Compute with Unlike Denominators," 78–80, 88; fractions with unlike denominators, 79–80; mixed numbers with unlike denominators, 80
"Spinning Equivalents," 73, 78
Square numbers, 24, 27
"Stop to Think About It," 59
"Strategies, Strategies, Strategies," 5, 6–8, 10
Strategy, 6

"Talking Problems," 116–117
"Talking Problems (Again)," 122
T-chart, 9, 36
Teaching tips: decimals, 99, 104, 107, 110, 115, 121; division, 33, 39, 42; fractions, 48, 52, 55, 60, 64, 69, 72, 76, 83, 87, 90, 94; multiplication, 2, 5, 13, 17, 20, 24, 28
"Third Shuffle," 66
"Traveling Computation," 38
"Trick of Tens," 14
"Two or More of the Same," 74

Unit fractions, 92

"Where's the One?", 72
"Why Bother?", 85; division-style, 92; division-style 2, 97

Zero Property of Multiplication, 6
"Zeros, Zeros Everywhere!", 41